THE COLLAPSE OF EVOLUTION

Scott M. Huse

Illustrated by

Janice Northcutt Huse

Reprinted 1986 by Baker Book House Company
with permission of copyright owner

ISBN: 0-8010-4310-7

Eighteenth printing, October 1993

All Scripture quotations are from the authorized
King James Version (KJV).

Printed in the United States of America

Dedication

This book is dedicated to our Creator and Saviour, the Lord Jesus Christ, Who alone is worthy to receive all glory, honor, and power.

CONTENTS

LIST OF FIGURES

FOREWORD

The theory of evolution has dominated our society for about a century, especially in our educational institutions. The media has been most influential in promoting the "fact" of organic evolution with some television programs and magazine editorial sections loyally devoted to the evolutionary viewpoint. Usually this indoctrination is obvious and insistent, but even when it is more subtle, it is nevertheless unmistakably effective.

Our current social and moral problems are largely a result of the humanistic philosophy which has been spawned by evolutionary thinking. The so-called "new morality" we are presently witnessing is actually "no morality," the inevitable result of the atheistic, evolutionary philosophy. Indeed, there are virtually no areas of thought and life today which have remained impervious to the effects of this popular viewpoint.

Ever since the Scopes Trial, evolution has been universally accepted as fact while at the same time Biblical creationism has been generally discarded as an antiquated belief of a former age. Organized Christianity has spent the past century or so retreating and compromising before the onslaughts of the evolutionists. Consequently, the faith of many Bible-believing Christians has been seriously affected and countless sincere people have been deceived.

Students all over the world are taught the historical and scientific greatness of the disproof of spontaneous generation (life coming from nonliving matter). Redi, Pasteur, and Spallanzani proved that life can only come from pre-existing life. Ideas such as mice coming from dirty undergarments were finally silenced. How ironic it is when these same educators turn right around and assert that spontaneous generation was the mechanism by which life arose! The modern concept of organic evolution, then, is actually nothing more than a refined regression to sixteenth century scientific mentality, where spontaneous generation is again proposed.

The irony of this whole situation is that **the very concept of organic evolution is completely absurd and impossible.** It is absolutely astonishing that an idea which is so devoid of any legitimate scientific evidence could have attained a position of such prestige in the name of science.

Biblical creationism, however, has recently been revived and is gathering impressive momentum. Literally thousands of distinguished scientists are rejecting evolution in favor of creationism. Many states are considering including creationism in the science curricula, and numerous lectures and debates have been hosted on university campuses.

The purpose of this book is basically three-fold: first, to expose the scientific fallacies of the theory of organic evolution; second, to present scientific evidence for Biblical creationism; and third, to prove that evolution and Biblical creationism are mutually exclusive and cannot be reconciled.

It has been said that there are two kinds of people: those who agree with you, and the bigots. There is no one so narrow as the poor misguided fellow who disagrees with you; and a logical person is anyone who can prove that you are right. The point is that to one degree or another we are all guilty of egocentrism. Although this is certainly true, I am nevertheless convinced that an honest seeker of the truth will eventually find it. With that in mind, I encourage the reader to proceed with an open mind and an honest heart.

"And ye shall seek me, and find me, when ye shall search for me with all your heart" (Jeremiah 29:13).

ACKNOWLEDGMENTS

The author wishes to express special appreciation to the following individuals and organizations who have helped with various aspects of the preparation of the manuscript: Illustrations—Janice Huse; Photography—NASA; Proof Reading and Manuscript Reviewing—Lisa Edwards, Rick Fineout, Preston Snowman, Bill Taylor, Steve Wilber; Layout and Design—Bruce Caisse; Typing—Lisa Edwards, Debbie Kiggins.

A special word of recognition is also due to Dr. Henry M. Morris, for many years the Head of the Department of Civil Engineering at Virginia Polytechnic Institute and now Director of the Creation Science Research Center, San Diego, California. Thanks to his original thinking and highly respected writings of the past decades, an impressive creation-science foundation has been established.

Finally, very special thanks are due to Bill Taylor who initiated and encouraged publication of this manuscript, and to my loving wife, whose patience and encouragement have made this project a reality.

INTRODUCTION

The most widespread and influential argument against the veracity of the Bible is the all too common belief that modern science has proven evolution, thereby discrediting the Scriptural account of Creation. The fatal flaw in this argument, however, is the fact that **it is impossible to prove scientifically any theory of origins.** This is because the very essence of the scientific method is based on *observation* and *experimentation,* and it is impossible to make observations or conduct experiments on the origin of the universe. This point is conceded by British biologist, L. Harrison Matthews, in the foreword to the 1971 edition of Darwin's *Origin of Species:*

> "The fact of evolution is the backbone of biology, and biology is thus in the peculiar position of being a science founded on an unproved theory — is it then a science or a faith? Belief in the theory of evolution is thus exactly parallel to belief in special creation — both are concepts which believers know to be true but neither, up to the present, has been capable of proof." 42

Scientists may speculate about the past or future but they can only actually observe the present. Obviously, then, **the widespread assumption that evolution is an established fact of science is absolutely false.** Thus, evolution can only be correctly labeled as a *belief,* a subjective philosophy of origins, the religion of many scientists. Despite this fact, most of today's scientists and teachers still insist that evolution is an established fact of science.

Sir Julian Huxley (grandson of Thomas Henry Huxley – Darwin's 'bulldog'), for example, pronounced in 1959 that:

> "The first point to make about Darwin's theory of evolution is that it is no longer a theory, but a fact. No serious scientist would deny the fact that evolution has occurred, just as he would not deny the fact that the Earth goes around the Sun." [70]

M.J. Kenny has stated:

> "Of the fact of organic evolution there can at the present day be no reasonable doubt; the evidences for it are so overwhelming that those who reject it can only be the victim's of ignorance or of prejudice." [12]

Professor George Gaylord Simpson assures us that:

> "Darwin . . . finally and definitely established evolution as a fact." [12]

Quoting from a standard geology textbook, *Essentials of Earth History,* we read:

> "The fossil record furnishes irrefutable proof that life on earth has changed through the ages . . . the systematic study of fossil remains cast an entirely new light on the past history of the earth and did away with the old-fashioned and superstitious notions on the subject that had prevailed for thousands of years . . . fossils prove not only that life has changed but also that it has progressed from simplicity to complexity with the passage of time. These are the facts. To those who take an unbiased view of the matter, there is only one conclusion – that all past and present life has descended from simple beginnings." [69]

Even the Pope's own Pontifical Academy of Sciences has recently stated:

> "We are convinced that masses of evidence render the application of the concept of evolution to man and the other primates beyond serious dispute."

While evolutionary scientists may differ about *how* evolution occurs, they do not disagree *that* it occurs. But we need to remember that scientists are human beings too. The notion that they are completely objective, detached, impartial, cold machines is, of course, absurd. The effects of prejudice and preconceived ideas, the influence of strong personal convictions, and the opinions of so-called "experts" influence them as much as anyone. Also, many scientists and teachers are unbelievers in Biblical Christianity. As such, they are forced to accept a naturalistic explanation for the origin and destiny of life and the universe. All of these factors play an extremely important role in the widespread acceptance of the evolutionary theory.

As Dr. George Wald, winner of the 1967 Nobel Peace Prize in Science, has written:

> "When it comes to the origin of life on this earth, there are only two possibilities: creation or spontaneous generation (evolution). There is no third way. Spontaneous generation was disproved 100 years ago, but that leads us only to one other conclusion: that of supernatural creation. We cannot accept that on philosophical grounds (personal reasons); therefore, we choose to believe the impossible: that life arose spontaneously by chance." 40

Thus, we find that evolution is generally accepted to be a fact of science, not because it can be proven by scientific evidence, but because the only alternative, special creation, is totally unacceptable.

Of course, not all scientists are atheistic evolutionists. In fact, some of the greatest pioneers in the history of science have been devoted Christians. A partial listing would include such names as Issac Newton, Louis Pasteur, Johannes Kepler, Robert Boyle, Michael Faraday, Samuel Morse, Lord Kelvin, and James Maxwell. (See Appendix C.)

Still other scientists have elected to favor a compromise position commonly known as *theistic evolution.* Theistic evolutionists claim to believe in God and the Bible, while at the same time, maintaining that all life has evolved from inorganic chemicals. But in accepting this curious concept, theistic evolutionists are forced to depart from, contradict, and compromise numerous Biblical essentials. Typical of such disregard for the Scriptures are the views of the well-known Jesuit priest, Pierre Teilhard de Chardin. However, Christians who truly believe the Bible to be the divinely inspired Word of God have no other recourse than to summarily dismiss and reject such a concept.

Simply stated, evolution may be defined as an imagined process by which living things formed by themselves without a creator and then somehow improved by themselves. According to this belief, all bacteria, plants, animals, and humans have arisen by mere chance from a single, remote ancestor that somehow came into existence. All of this is supposed to have occurred *accidentally* without the benefit of any intelligence or planning. The basic premise of this "molecule-to-man" theory is that hydrogen gas, given enough time, will eventually turn into people. Diametrically opposed to this viewpoint, Biblical creationism postulates an initial special creation by God through which all the laws, processes, and entities of nature were brought into existence as described in the book of Genesis.

An important point worthy of contemplation is the

fact that throughout history every age has been beset with false ideas. For example, it was believed for 15 centuries that the sun and other planetary bodies revolved around the earth. This idea was known as the *geocentric theory* and, of course, we now know this supposition to be completely false. During the 17th and 18th centuries another theory was universally accepted and taught as an established fact of science in much the same way that evolution is today. This theory was known as the *phlogiston theory.* It stated that every substance which burned contained the magic ingredient *phlogiston,* which gave it combustibility. This erroneous idea was later refuted by the French chemist, Lavoisier, who demonstrated conclusively that *oxygen* was the key element involved in combustion. [77]

There are many other historical blunders of science which could be mentioned, but these should suffice to establish the fact that every age has been plagued with error. Are we to presume that we are any different or exempt from such folly? Is it possible that the theory of evolution is yet another blunder of science? In this book we shall endeavor to demonstrate that this is indeed the case.

But why bother to study origins anyway? Is it really worth all the time and energy? Well, there are many good reasons why it is important to have a correct understanding of origins, and yes, it is a worthwhile pursuit. Everyone needs a sense of identity, purpose, and personal goals. This is impossible without a sense of origin. What a person believes about his or her origin will condition that person's life-style and affect his or her ultimate destiny.

The solutions to man's massive social problems depend upon a correct understanding of origins. If the evolutionary philosophy is correct, then life is without moral direction and purpose. On the other hand, if we were created by God, our lives have meaning, direction, and purpose. Clearly, the proposition of origins is the foundation of all other convictions, actions and beliefs. Thus, the question of origins is a **vital issue** that can be ignored only at great peril.

Although both evolution and Biblical creationism are inaccessible by the scientific method and must be accepted by faith, this is not to say that known scientific facts and observations are useless and to be discarded or ignored. On the contrary, scientific facts should be used to help ascertain which model of origins is more probable so that the invested faith is intelligent and not blind. In this book we shall endeavor to demonstrate that the facts of science correlate much better with Biblical creationism than with evolution, and that evolutionists are indeed in a most precarious position when sitting on the chair of the general theory of evolution, as expressed in the following poem:

> "As I was sitting in my chair,
> I knew it had no bottom there,
> Nor legs, or back, but I just sat,
> Ignoring little things like that."

1.

GEOLOGY

A. The Geologic Column and Timetable

1. Introduction

Before the nineteenth century the vast majority of scientists interpreted earth history in terms of Biblical creationism and catastrophism (Genesis Flood), and consequently, believed in a relatively short time scale. However, the more recent acceptance of a principle known as *uniformitarianism* has successfully promoted the idea of an ancient earth. Uniformitarianism is the belief that the origin and development of all things can be explained exclusively in terms of the same natural laws and processes operating today. According to this dogma, nature can be satisfactorily explained according to natural causes and therefore, "the present is the key to the past." This concept was introduced by James Hutton, popularized by Sir Charles Lyell, and greatly influenced the thoughts and works of Charles Darwin. Uniformitarianism has been the backbone of modern historical geology and is responsible for the current widespread assumption that the earth is billions of years old. Consequently, the earth has *aged* from just a few thousand years to about 5 billion years in a little more than a century!

Present-day uniformitarian geologists generally evince a hostile attitude toward Biblical creationism and catastrophism. L. Merson Davies, a renowned British paleontologist who opposed the general theory of evolution, commented on this recent phenomenon and made some important observations:

> "Here, then, we come face to face with a circumstance which cannot be ignored in dealing with this subject . . . namely, the existence of a marked prejudice against the acceptance of belief in a cataclysm like the Deluge. Now we should remember that, up to a hundred years ago, such a prejudice did not exist . . . as a general one, at least. Belief in the Deluge of Noah was axiomatic, not only in the Church itself (both Catholic and Protestant) but in the scientific world as well. And yet the Bible stood committed to the prophecy that, in what it calls the "last days," a very different philosophy would be found in the ascendent; a philosophy which would lead men to regard belief in the Flood with disfavor, and treat it as disproved, declaring that "All things continue as from the beginning of the creation" (2 Peter 3:3-6). In other words, a doctrine of Uniformity in all things (a doctrine which the apostle obviously regarded as untrue to fact) was to replace belief in such cataclysms as the Deluge." [21]

Thus, with the doctrine of uniformitarianism, Peter's ancient prophecy has at last been fulfilled before our own eyes. According to modern uniformitarians, there is no need to rely on any catastrophic events, except on a minor scale. They insist that all geologic features and formations, once attributed to geologic cataclysms, can now be satisfactorily explained by ordinary processes functioning over long periods of time. Although the uniformitarian assumption appears to be reasonable, we must remember that it is merely an assumption and not a fact. Furthermore, it is an assumption which has been made in the construction of the geologic column, and in the pages that follow we intend to demonstrate that it is an inadequate mechanism in terms of explaining most geologic features and formations. Instead, it will be shown that the geologic evidences support a universal hydraulic cataclysm. While uniformitarian evolutionists insist that *the present is the key to the past,* creationists argue that it is actually *the past* (the fall and redemption of man) *that is the key to the present* (the problems of the world today).

2. The Geologic Column

Among the sciences, historical geology is in a most awkward position in that it deals with past events and is thus forced to rely upon assumptions which may or may not be true. Documented history only goes back a few thousand years. The earliest authenticated written records date back to about 3,500 B.C. [54] Prior to the existence of eyewitnesses, no one can be absolutely certain of what actually transpired. Consequently, there is no direct irrefutable proof as to the process(es) which formed the rocks of the geologic column or their age. Any such determination can only be indirect, based upon assumptions which may or may not be true.

Nevertheless, based upon the assumptions that uniformitarianism and organic evolution were established scientific facts, geologists during the 19th century, began to compile the geologic column (Figure 1). They arranged the earth's strata according to the various types of fossils they contained, especially their *index fossils* (usually marine invertebrates which are easily recognized, assumed to have been widespread in occurrence, and of limited chronological duration thus marking a specific age determination for a rock formation; see Figure 2). Strata with simpler fossils (*presumed to have evolved first*) were put on the bottom of the column while strata containing more complex forms (*presumed to have evolved later*) were placed toward the top of the

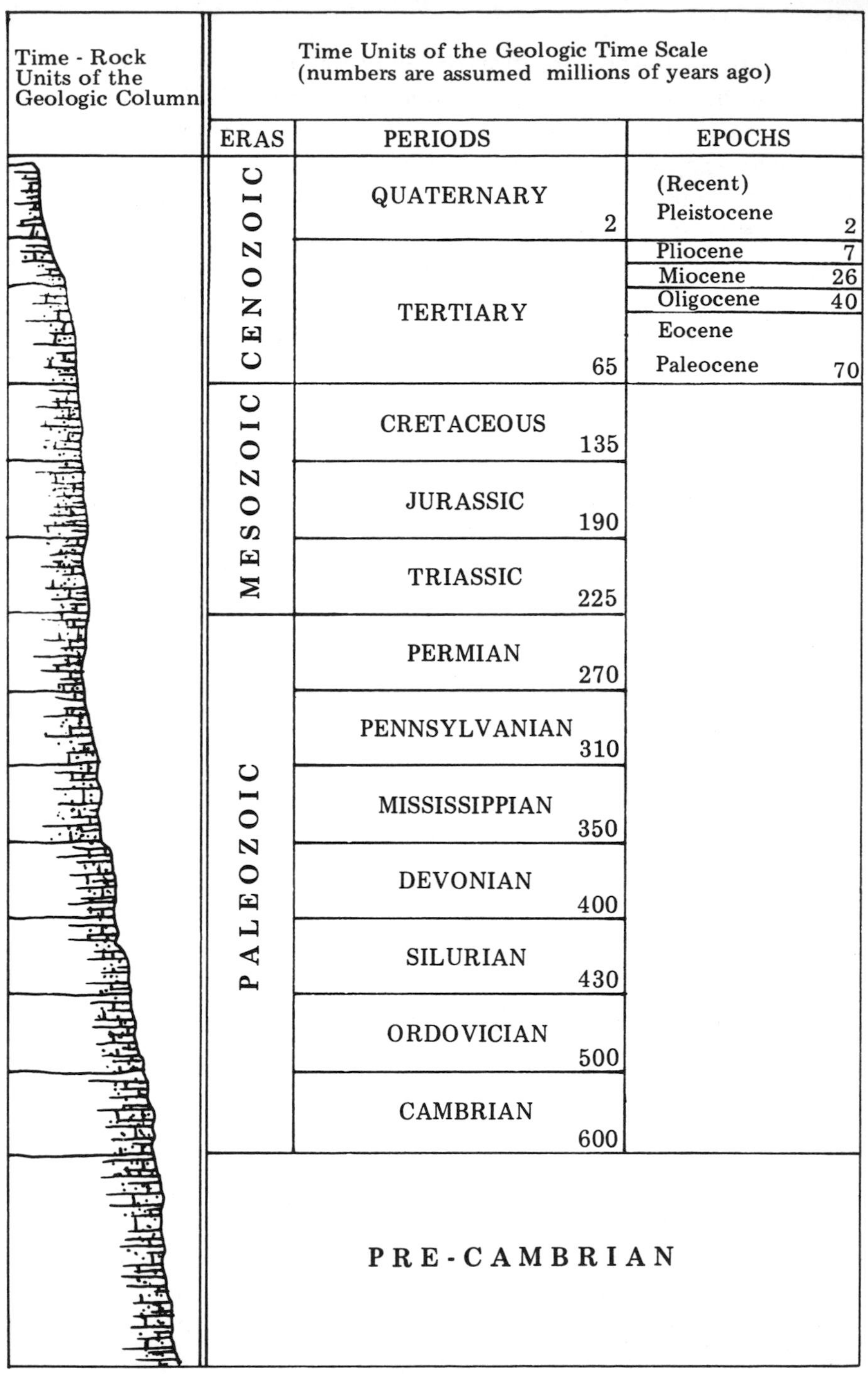

FIG. 1: The Geologic Column and Timetable. The Geologic Column, as shown here in its ideal, continuous sequence, does not exist anywhere in the world.

Characteristic Life According to Evolutionary Model	Possible Characteristics According to Creation Model.	Relative Lengths of Geologic Eras
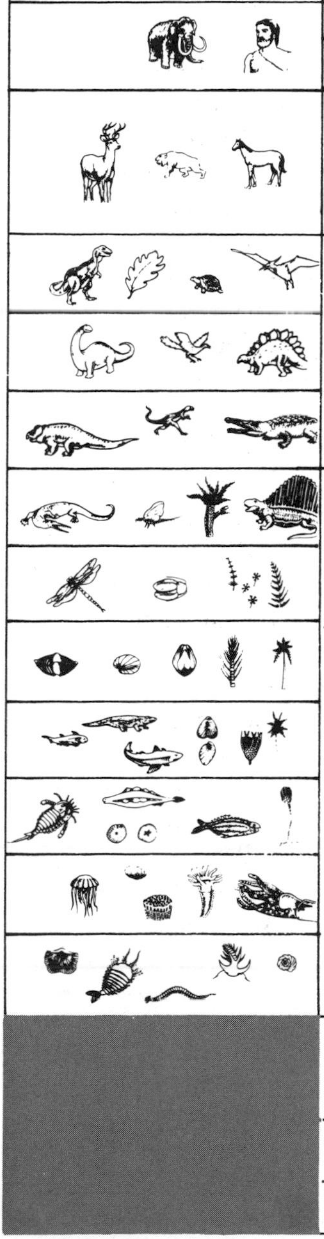	Post-flood world characterized by modern processes.	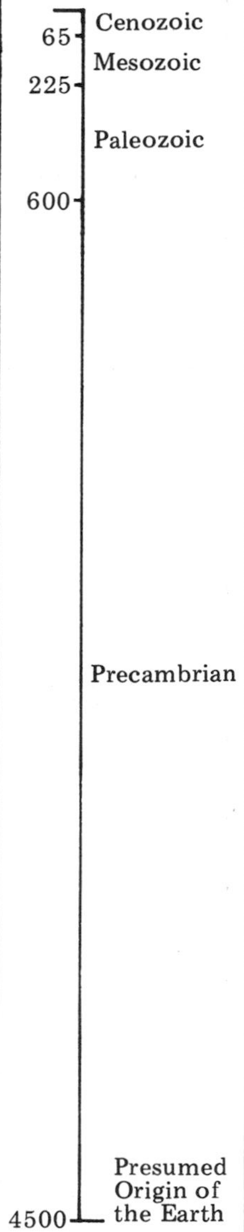
	Ice Age. Effects of post-flood glaciation and pluviation. Volcanism and tectonism subsiding.	65 Cenozoic
	Final stages of flood and early post-flood activity. Water drains into basins eroding previously deposited sediments.	Mesozoic 225 Paleozoic 600
	Intermediate stages of flood, with mixtures of marine and continental deposits. Land submerged completely. Extinction of all land creatures, including dinosaurs. Oceans begin to deepen or widen at end of this phase.	
	Waters nearing maximum height. Primarily upper marine and shelf-type deposits and fossils. Some mixing with land plants and animals.	
	Mats of vegatation floating on open sea, as forests are uprooted.	Precambrian
	Deposits cover greater area as water level rises. Fossils still mostly marine.	
	Early-flood phases; sedimentation mostly in pre-flood basins. Ocean bottom dwellers trapped in deep-sea deposits.	
	Mostly pre-flood marine sedimentation. Simple, non-mobile fossils trapped in quiet waters between the curse and the flood.	
	Fossil-free sedimentary rock dating from before the creation of life.	4500 Presumed Origin of the Earth
	Origin of crust dating from the first day of the creation week.	

Included in this figure is one possible interpretation of the Geologic Systems according to the creation model.

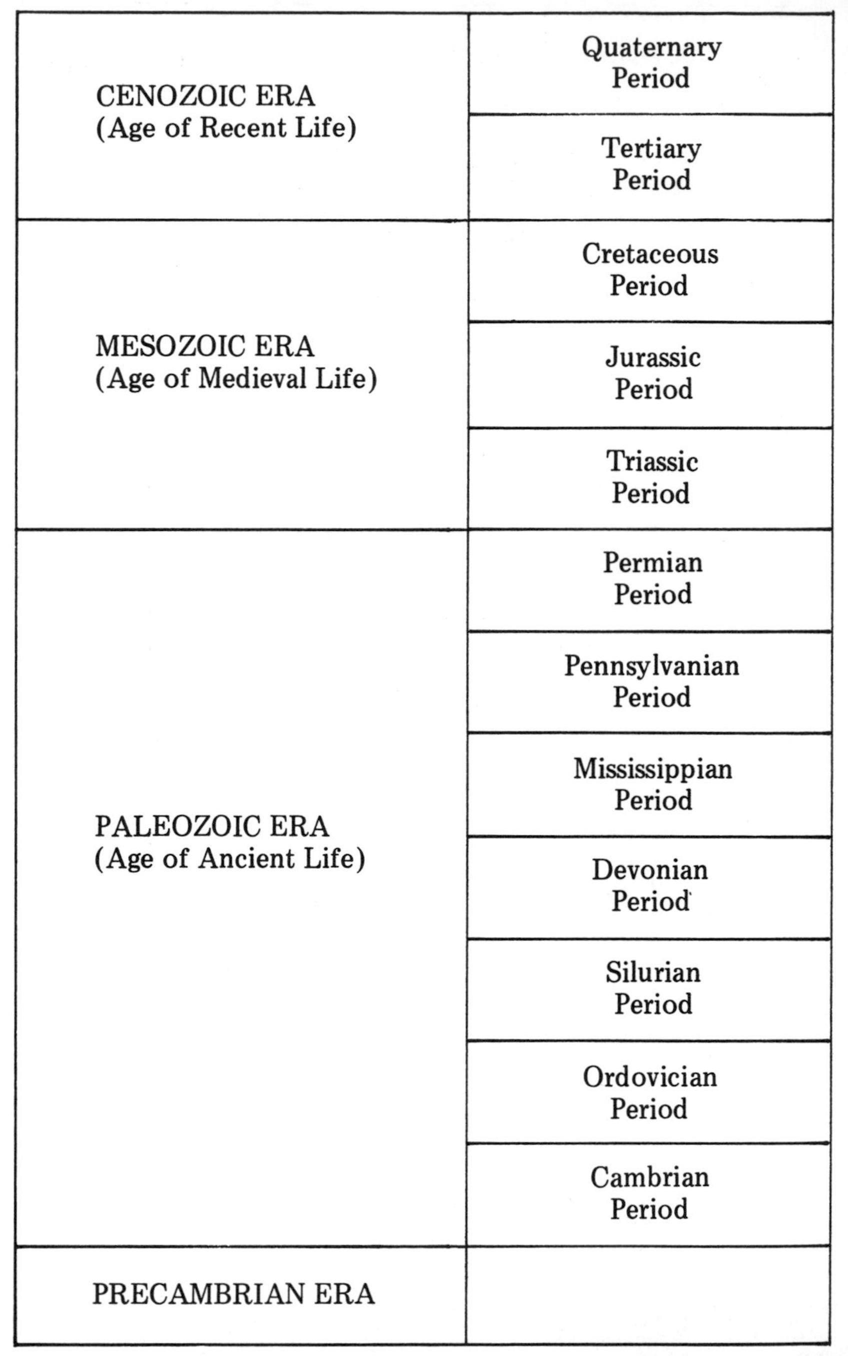

CENOZOIC ERA (Age of Recent Life)	Quaternary Period
	Tertiary Period
MESOZOIC ERA (Age of Medieval Life)	Cretaceous Period
	Jurassic Period
	Triassic Period
PALEOZOIC ERA (Age of Ancient Life)	Permian Period
	Pennsylvanian Period
	Mississippian Period
	Devonian Period
	Silurian Period
	Ordovician Period
	Cambrian Period
PRECAMBRIAN ERA	

FIG. 2: Index Fossils. Index fossils are forms of life which are presumed to have existed during limited periods of geologic time and to have been widespread in occurrence thus indexing the age of the rocks in which

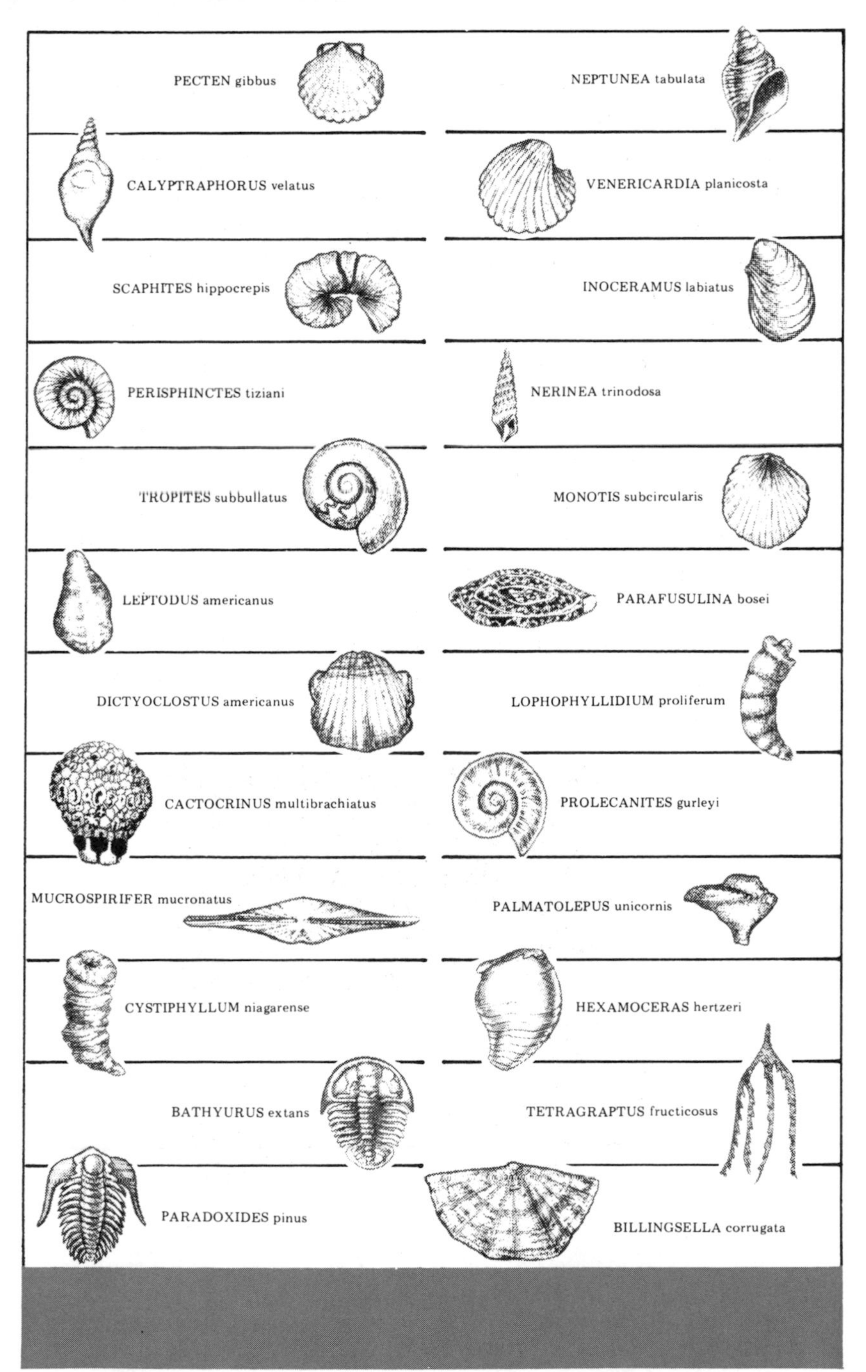

they are found. Creationists maintain that such fossil zonation does not indicate evolution but reflects the combined effects of hydraulic, ecologic, and physiologic sorting in conjunction with the Genesis Flood.

column. Thus, **the entire geologic column was founded and built on the assumption that organic evolution was a fact.** The fact that modern historical geology is based on the assumption of evolutionary biology is a blatant case of circular reasoning (Figure 3). The only basis for placing rock formations in chronological order is their fossil content, especially index fossils. The only justification for assigning fossils to specific time periods in that chronology is the assumed evolutionary progression of life. In turn, the only basis for biological evolution is the fossil record so constructed. In other words, the assumption of evolution is used to arrange the sequence of fossils, then the resultant sequence is advanced as proof of evolution. Consequently, **the primary evidence for evolution is the assumption of evolution!** Since the arrangement of fossils is completely arbitrary (i.e., based on the assumption of evolution) the geologic column cannot be used to demonstrate evolution or a vast geologic age. [35]

FIG. 3: Circular Reasoning. Historical geology is based on the assumption of evolutionary biology. Thus, the primary evidence for evolution is the assumption of evolution! This is a blatant case of circular reasoning that is completely invalid.

It is important to realize that **nowhere in the world does the geologic column actually occur.** It exists only in the minds of evolutionary geologists. It is simply an idea, an ideal series of geologic systems, and not an actual column of rocks that can be observed at a particular locality. Real rock formations are characterized by gaps and reversals of this ideal, imaginary sequence. Even the Grand Canyon only includes less than half of the geologic systems. [8] In order to see the entire geologic column as it occurs in its "proper" sequential order, one would have to travel all over the world. Pre-Cambrian and Paleozoic strata can be seen at the Grand Canyon; Mesozoic strata can be observed in eastern Arizona; Tertiary formations are visible in New Mexico, and so on.

In the field every conceivable contradiction to the proposed ideal sequence of the geologic column is found. Anomalies such as fossils in the *wrong* stratigraphic order must be explained away if the basic assumption of evolution and the validity of the geologic column are to be salvaged. The evolutionary explanation for this phenomenon is as clever as it is convenient. Fossils found *too low* in the geologic column (i.e., before they were supposed to have evolved) are termed *stratigraphic leaks.* Specimens found *too high* in the geologic column are considered as *re-worked* specimens. [60] In many cases huge thicknesses of sedimentary strata are found to be in the "wrong" stratigraphic order. Evolutionary geologists resort to secondary assumptions in order to explain such upside-down conditions (older strata on top of younger strata). Overthrusting is one such imaginative concept which is used to account for such anomalies. For example, in Glacier National Park there is a block of Pre-Cambrian limestone (supposedly 1 billion years old) on top of a Cretaceous shale formation (supposedly only 100 million years old). Geologists propose that this "misplaced" block has been moved horizontally over the adjacent region driven by tremendous forces along thrust faults. But this "misplaced" block of limestone is about 350 miles long, 35 miles wide, and 6 miles thick! Dr. Henry M. Morris has demonstrated conclusively that such large-scale mechanical sliding would be physically impossible even if the sliding planes were lubricated. The obvious conclusion is that the *Pre-Cambrian rocks* were actually deposited after the *Cretaceous rocks* and that the geologic column and timetable of earth history is basically meaningless. Such concepts as large-scale overthrusting are simply viewed by creationists as ploys to explain away the numerous contradictions encountered in the field.[73]

The fact that the rock formations of the geologic column are arranged and dated by their fossil contents (especially their index fossil content) and not on the basis of radiometric dating techniques surprises most people. This common misconception is, however, easily dispelled by the simple fact that long before radioactive dating was ever imagined the geologic column and approximate ages of all fossil-bearing strata were already established. Radiometric dates, which are estimated on certain rock formations, are usually discarded and never used at all due to the high probability of error inherent in this technique. This is especially true whenever such determinations disagree with the previously established dates.[32]

3. Recent Findings Which Contradict the Geologic Column

There are a number of known scientific facts which raise serious questions concerning the geologic column and timetable. One such example would be the existence of numerous contemporaneous human and dinosaur prints found in Mexico, New Mexico, Arizona, Missouri, Kentucky, Illinois, and in other U.S. localities. These tracks are widely distributed and are usually only exposed by flood erosion or bulldozers. They have been carefully studied and verified by reliable paleontologists and cannot be dismissed as frauds. [78] Furthermore, there are places in Arizona and Rhodesia where dinosaur pictographs have been found drawn on cave or canyon walls by man.[50] The obvious implication is that man once lived contemporaneously with dinosaurs, contrary to the commonly accepted chronology of the geologic column and timetable. Job 40:15–41:34 is an interesting reference in this connection which seems to refer to land and marine dinosaurs living in Job's day.

Five-toed llamas allegedly became extinct about 30 million years ago according to the evolutionary framework. Yet, archaeologists have found pottery with etchings of five-toed llamas on it. Skeletons of five-toed llamas have also been found in association with the Tiahuanacan culture. [16]

An ancient Mayan relief sculpture of a bird resembling the Archaeopteryx has been found. This indicates a discrepancy of about 130 million years. If the geologic column is correct, the two should never have met. Apparently, the geological column is in error. [78]

An amazing discovery was made by William Meister on June 1, 1968 in Utah. He found the fossils of several trilobites in the fossilized, sandaled footprint of a man! But according to the evolutionary timetable worked out in the geologic column, trilobites became extinct about 230 million years before the appearance of man! Thus, to find a modern, sandal-shod man existing contemporaneously with trilobites is utterly devastating to the geologic column and evolutionary framework.[17]

Dr. Clifford Burdick, a geologist, found additional evidence to support the hypothesis that man did indeed once live contemporaneously with trilobites. He discovered the footprints of a barefooted child, one of which contained a compressed trilobite.[75]

Naturally, evolutionary geologists have reacted to this evidence in a rather skeptical and evasive manner. If they accept this evidence, the credibility of the geologic column and timetable, which has been a dogma of geology for generations, would be shattered. Yet, the evidence is real and cannot be denied or ignored any longer. Many other contradictions could be listed which raise serious questions concerning classical evolutionary thinking in the geologic column and timetable. For additional information concerning such contradictions see *Time Upside Down* by Erich von Fange (1974).

4. Radioactive Dating

The belief in an ancient earth has become so entrenched in our modern culture that advocates of a relatively young earth are considered as antiquated as if they had proposed that the earth were flat. In fact, it may even seem audacious to question the validity of radioactive dating. After all, haven't scientists proven that the earth is billions of years old by these very methods?

As was established earlier, the study of origins is beyond the reach of the scientific method. Therefore dates obtained from these techniques are merely circumstantial and are necessarily based on numerous *assumptions,* which may or may not be true. Thus, **it is impossible to prove that the earth is billions of years old.**

Radioactive dating techniques may be classified into two main categories: (1) Those whose information is limited to the last few thousand years, such as carbon-14; and (2) those which utilize radioactive elements such as uranium-lead and potassium-argon for determining dates supposedly stretching back into the millions and billions of years.

In dating the earth, scientists rely ,primarily on the uranium-lead and potassium-argon methods. Of these two, the uranium-lead method is more important because it is the one against which the others have been calibrated. Consequently, it is the method which offers the primary support for the belief of an ancient earth held to be around 4.5 to 5 billion years old. Regardless of the particular method utilized, dates obtained from these techniques are indeed circumstantial and based on a number of assumptions which we will now consider.

When using radioactive dating techniques, the following assumptions must be made:

(1) the rock contained no daughter product atoms in the beginning, only parent atoms;

(2) since then, no parent or daughter atoms were either added to, or taken from the rock; and

(3) the rate of radioactive decay has remained constant.

Depending on the particular method involved, other assumptions may be involved, but these three are always involved and are extremely important. Recognizing this fact, the dubious nature of radiometric dating techniques now becomes apparent especially since *none* of these assumptions are found to be valid!

First of all, there is no way to be absolutely certain that some daughter product atoms were not present in the rock in the beginning since they are all found to be widely distributed in the earth's crust. Secondly, heating and deforming of rocks can cause migration of the daughter and parent atoms. Percolation of water through the rocks can also cause these atoms to be transported and redeposited elsewhere. And finally, recent research suggests that certain conditions (exposure to neutrino, neutron, or cosmic radiation) may alter the rates of radioactive decay.[36] Radioactive decay is also known to be proportional to the speed of light, and Barry Setterfield has recently shown that the speed of light has not remained fixed but has actually decreased. This decrease in the speed of light suggests that the decay of radioactive material in the rocks in the past would be much greater than it is today. Thus, the high decay rates of the past would account for the apparent vast age of the rocks. [11]

The fact that erroneous results can be and often are derived from these dating techniques has been experimentally verified. For example, living snails have been dated as 2,300 years old by the carbon-14 method.[34] Wood taken from growing trees has been dated by the carbon-14 method to be 10,000 years old.[31] Hawaiian lava flows, which are known to be less than 200 years old, have been dated by the potassium-argon method at up to 3 billion years old![37] These scientists would do well to ponder the question that God put to Job: "Where wast thou when I laid the foundations of the earth?" (Job 38:4a).

Considering the unreliability of these dating techniques, there is no valid scientific reason why one should place any confidence in them especially since there is such an immense amount of extrapolation involved. Consequently, we will now consider other processes which can also be used to measure time and thus date the earth.

5. Processes Which Indicate a Young Earth

A significant amount of research has been accumulated which suggests that the earth and solar system are relatively young and not ancient as evolutionists have assumed. The following processes are usually selectively screened out by evolutionists, who choose to ignore or discard them, because they indicate a relatively young age for the earth and solar system.

(a) The Earth's Magnetic Field

The strength of the earth's magnetic field has been measured for well over a century. This provides scientists with exceptionally good records. In an important recent study, Dr. Thomas G. Barnes has shown that the strength of the earth's magnetic field is decaying exponentially at a rate corresponding to a half-life of 1,400 years (Figure 4). That is to say, 1,400 years ago the magnetic field of the earth was twice as strong as it is now. If we extrapolate back as far as 10,000 years, we find that the earth would have had a magnetic field as strong as that of a magnetic star! This is, of course, highly improbable, if not impossible. Thus, based on the present decay rate of the earth's magnetic field, 10,000 years appears to be an upper limit for the age of the earth.

Keep in mind that **any objections to this conclusion must be based on rejection of the same uniformitarian assumption which evolutionists utilize to derive a great age for the earth.** [54]

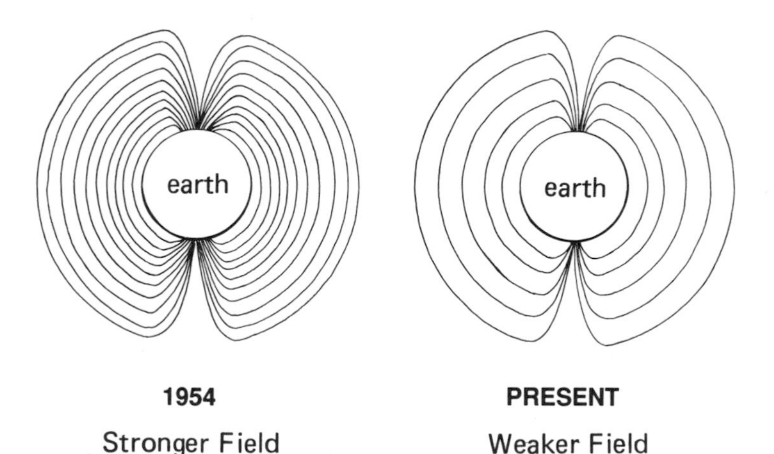

FIG. 4: The Earth's Magnetic Field. The strength of the earth's magnetic field has been continuously weakening. It has been decaying exponentially at a rate corresponding to a half-life of 1,400 years.

In defense of their long-age chronology, evolutionists have proposed a *reversal hypothesis.* They suggest that the earth's magnetic field has remained relatively stable throughout geologic time, except for certain intervals in which it went through a reversal, dying down to zero and rising up again with the reverse polarity. The last such reversal is alleged to have occurred about 700,000 years ago.

Unfortunately for evolutionary scientists, the reversal hypothesis has absolutely no valid scientific theoretical basis. Furthermore, rock magnetization cannot be used to support these so-called *reversals* because there is a self-reversal process known to exist in rocks, completely independent of the earth's magnetic field.[4]

Finally, it is believed that the earth's magnetic field is due to circulating electric currents in its core. If we extrapolate backward about 20,000 years, we find that the estimated heat produced by the currents would have melted the earth.[3] Clearly, the testimony of the earth's magnetic field is strongly in favor of a relatively young earth, not an ancient one.

(b) Meteoritic Dust

Scientists have known for some time now that cosmic dust particles enter the earth's atmosphere from space at an essentially constant rate. Eventually these dust particles settle down to the earth's surface. Hans Petterson has made accurate measurements of this influx, and has determined that the earth receives about 14 million tons per year. Now, if it is true that the earth is around 5 billion years old, as evolutionists insist, there should be a layer of meteoritic dust that is about 182 feet thick all over the world! No such dust layer exists anywhere, of course. Even on the moon where it should be at least this thick, little sign of it was found by the astronauts (about an eighth of an inch). The fear that the astronauts would sink into such a layer of meteoritic dust when they landed on the moon proved to be completely unwarranted (Figure 5).

Evolutionary geologists might be tempted to argue that erosional and mixing processes can account for the absence of the meteoritic dust layer. However, this explanation is unsatisfactory and easily refuted as the composition of meteoritic dust is very distinctive, particularly in its content of nickel and iron. For example, nickel is a very rare element in the earth's crust and even more scarce in the ocean. But the average nickel content of meteoritic dust is approximately 300 times as great as in the earth's crust! Calculations based on the relatively small amount of nickel found in the earth's crust and ocean indicate an age for the earth of only a few thousand years.[54]

(c) The Mississippi River Delta

The Mississippi River delta offers additional evidence to support the concept of a relatively young earth. Approximately 300 million cubic yards of sediment are deposited into the Gulf of Mexico by the Mississippi River each year. By carefully studying the volume and rate of accumulation of the Mississippi River delta and then dividing the weight of the sediments deposited annually into the total weight of the delta, it

FIG. 5: The Lunar Surface. Astronaut Edwin Aldrin walking near the Lunar Module Eagle. The footprints in the foreground reveal the reality of a modest dust layer on the surface of the moon. (NASA)

can be determined that the age of the delta is about 4,000 years old.[78]

(d) Petroleum and Natural Gas

Petroleum and natural gas are contained at high pressures in underground reservoirs by relatively impermeable cap rock. In many cases, the pressures are extremely high. Calculations based upon the measured permeability of the cap rock reveal that the oil and gas pressures could not be maintained for much longer than 10,000 years in many instances. Thus, the generalization that such fossil-fuel deposits have been confined for millions of years, having not leaked out through their cap rock, becomes preposterous.[36]

Furthermore, recent experiments have demonstrated conclusively that the conversion of marine and vegetal matter into oil and gas can be achieved in a surprisingly short time. For example, plant-derived material has been converted into a good grade of petroleum in as little as 20 minutes under the proper temperature and pressure conditions. Wood and other cellulosic material have also been converted into coal or coal-like substances in just a few hours. These experiments prove that the formation of coal, oil, and gas did not necessarily require millions of years to form as uniformitarian geologists have assumed and taught.[49]

Creationists believe that the great coal deposits of the world are the transported and metamorphosed remains of the extensive vegetation of the antediluvian world. This catastrophic interpretation is further supported by the presence of polystrate fossils in coal beds which indicate rapid formation. Also, the type of plants involved and the texture of these deposits testify of turbulent waters, not a stagnant swamp.[60]

Evolutionists propose that coal was formed millions of years before man evolved. However, human skeletons and artifacts, such as intricately structured gold chains, have been found in coal deposits. In Genesis 4 we learn that metal working was already highly developed; Tubal-cain was an instructor of every artificer in brass and iron. In Genesis 7 and 8, the Deluge buried the antediluvian civilizations in the sedimentary layers of the earth's crust.

(e) The Rotation of the Earth

The rotation of the earth is gradually slowing due to the gravitational drag forces of the sun, moon, and other factors. If the earth is billions of years old, as uniformitarian geologists insist, and it has been slowing down uniformly, then its present rotation should be zero! Furthermore, if we extrapolate backwards for several billion years, the centrifugal force would have been so great that the continents would have been sent to the equatorial regions and the overall shape of the earth would have been a flat pancake. But, as is commonly known, the shape of the earth is spherical; its continents are not confined to the equatorial regions, and it continues to rotate on its axis at approximately 1,000 miles per hour at the equator. The obvious conclusion is that the earth is not billions of years old.[78]

(f) The Recession of the Moon

A very simple proof that the earth and moon are relatively young is found in the recession of the moon from the earth. The present rate of recession of the moon is known and clearly indicates a young age for the earth-moon system.

The basic problem with which evolutionists have to contend is that the moon is presently much too close to the earth. Calculations based on the known recession speed of the moon and presumed evolutionary age of 4 to 5 billion years require that the moon should be much farther away from the earth than it is. Obviously, the earth-moon system is not as old as evolutionary scientists have assumed. The vast time span essential for the presumed evolution of life forms is apparently mythological and non-existent.[6]

(g) Atmospheric Helium

Another excellent evidence for a young earth is provided by the small amount of helium in our present-day atmosphere. Evolutionists maintain that the radioactive decay processes of uranium and thorium that produce helium have been occurring in the earth's crust for billions of years. But if this decay has been going on for billions of years, the earth's atmosphere should contain much more than the present 1 part in 200,000 of helium. The common explanation offered for the absence of the required helium is that it has been escaping out through the exosphere. But there is no evidence to support this assumption and, in fact, recent data indicate that helium cannot escape into space the way hydrogen does. To make matters worse for the somewhat discouraged evolutionist, it is also probable that helium is actually entering the atmosphere from outer space by means of the sun's corona. Realistic calculations based on available figures disclose that the amount of time required for natural alpha decay processes to have produced the observed helium is approximately 10,000 years.[77, 78]

(h) Pleochroic Halos

Creationists point to the evidence from study of rapidly decaying radioactive elements and pleochroic halos to support their belief that creation was sudden and complete. Dr. Robert Gentry, the world's leading authority on radiohalos, has studied the mysterious case of polonium radiohalos extensively and reached some startling conclusions which confirm the creationist's viewpoint.

Polonium 218 has been considered a daughter element of the natural decay of uranium, but through the works of Dr. Gentry and other researchers, polonium halos have been found in mica and fluorite without any evidence of *parents.* In other words, it was primordial—present in the original granite from the very beginning. Also, and most significantly, polonium halos should not exist at all because of their extremely short half-lives. Polonium 218 has a half-life of only 3 minutes. If the evolutionist's interpretation was correct and the rock formations gradually cooled over millions of years, the polonium would have decayed into other elements long ago.

Thus, the evidence clearly points to an instantaneous crystallization of the host basement rocks of the earth concurrent with the formation of the polonium. Simply stated, the presence of polonium radiohalos is one of the greatest blows to evolutionary thinking because it speaks so eloquently of instantaneous creation. [43]

While these facts raise absolute havoc with the evolutionary framework, they are in complete harmony with the creationist's viewpoint. As the Scriptures record: ". . . the evening and the morning were the first day" (Genesis 1:5). The deliberate and emphatic repetition of this phrase throughout Genesis 1 clearly indicates that these were literal 24-hour creative periods. The creation was accomplished in 6 literal days, not through billions of years of gradual development (Exodus 20:11). How appropriate are the words of the Psalmist at this juncture:

> "By the word of the Lord were the heavens made; and all the host of them by the breath of His mouth. For He spake, and it was done; He commanded, and it stood fast."
>
> (Psalm 33:6, 9)

(i) Population Growth

Another process that offers convincing evidence for a relatively young earth is that of population growth. Evolutionists believe that man has been on the earth for at least one million years, whereas creationists believe that he has been around for only a few thousand years. The question then becomes, "which possibility is better supported by the data from population growth statistics?"

Dr. Henry M. Morris has calculated that an average growth of only ½% per year, which is ¼ the present rate, would yield the present population in just 4,000 years. This allows ample room for periods of time when, because of war or disease, the population growth rates were far below the normal averages. Dr. Morris points out that it is statistically inconceivable that only 3.5 billion people could have resulted from one million years of evolutionary history. Even if the population increased at only ½% per year for a million years, the number of people in the present generation would exceed 10^{2100}! To fully appreciate the ludicrous nature of the evolutionary model in this regard, consider the fact that only 10^{130} electrons can be packed into the entire universe! Talk about crowded! Obviously, the creation model of human chronology offers the more reasonable figures on man's antiquity. Man's history clearly spans only thousands of years, not millions.[54]

6. Processes Which Indicate a Young Solar System and Universe

(a) Comets

Comets journey around the sun and are assumed to be the same age as the solar system. Each time a comet orbits the sun, a small part of its mass is "boiled off" (Figure 6). Careful studies indicate that the effect of this dissolution process on short-term comets would have totally dissipated them in about 10,000 years. Based on the fact that there are still numerous comets orbiting the sun with no source of new comets known to exist, we can deduce that our solar system can not be much older than 10,000 years. To date, no satisfactory explanation has been given to discredit this evidence for a youthful solar system. [36]

(b) Poynting-Robertson Effect

The sun, acting like a giant vacuum cleaner, sweeps up about 100,000 tons of micrometeoroids each day. The sun's radiation pressure also serves to push small inter-planetary dust particles into space. This phenomenon is known as the *Poynting-Robertson effect.* If the solar system is truly billions of years old, these particles should no longer be present. Proceeding at its present rate, the sun would have *cleaned house* in less than 10,000 years as there is no known source of appreciable replenishment. However, micrometeoroids are copious throughout the solar system, and this fact speaks convincingly for a relatively young solar system. [78]

(c) Star Clusters

Star clusters serve to indicate a young age for the universe. A star cluster contains hundreds or

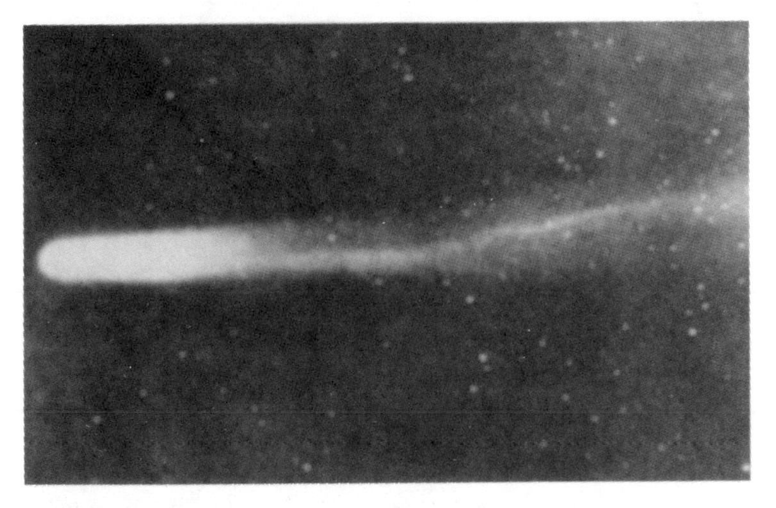

FIG. 6: Halley's Comet. Halley's comet is just one of many comets present in our solar system and should not exist if the solar system is as old as evolutionary scientists insist.

thousands of stars moving, as one author has put it, "like a swarm of bees" (Figure 7). They are held together by gravity, but in some star clusters, the stars are moving so fast that they could not have held together for millions or billions of years. Thus, the presence of star clusters in the universe indicates that the age of the universe is numbered in the thousands of years. [36]

(d) Super Stars

The energy given off by our sun has been computed to equal that of a billion hydrogen bombs being detonated every second. Some stars are so large and bright that they radiate energy anywhere from 100,000 to 1 million times as fast as our own sun! These stars could not have contained enough hydrogen to run the atomic fusion energy production process at such rates for millions or billions of years because their initial mass would have been absolutely implausible. Therefore, these stars must not be billions or even millions of years old, but rather only thousands of years old. [36]

The aforementioned examples are but a mere sampling

FIG. 7: A Star Cluster. Thousands of gravitationally bound stars are here shown in the globular star cluster in Hercules (M13), on the edge of the Milky Way Galaxy.

of the many processes which indicate a relatively young age for the earth, solar system, and universe. Numerous other examples could be cited in this context, but these will suffice to establish the point that there is considerable and convincing evidence to support the concept of youthful universe, contrary to popular evolutionary thinking. An important point in this connection is the fact that implicit in Biblical creation is the appearance of an age beyond the actual age. In other words, a supernatural creation would give the appearance of age beyond that which would normally be required for the same event to occur naturally. For example, the bread, fish, and wine, which Jesus miraculously created, gave a sudden appearance of age. Thus, creation may well have been mature from its inception, giving only an appearance of age.

It is important to remember that just because most people believe in something does not necessarily make it true. The argument from a majority opinion is not an impressive one. In fact, it is completely irrelevant since scientific truth is never determined by taking a vote. Majorities can be, and often have been, completely wrong.

7. The Creationist's Interpretation

In view of the numerous fallacies of the evolutionary model and the abundance of scientific facts that support Biblical creationism, we shall now consider the Biblical creationist's interpretation. All Bible-believing creationists agree that God is the Creator. Admittedly, however, not all creationists agree as to the exact method of His creation. There is considerable diversity of opinion among creationists regarding the interpretation and application of Genesis 1 to earth history. Some Bible "scholars" have attempted to harmonize the Genesis account of creation with the evolutionary framework of earth history; others prefer to remain loyal to the more traditional viewpoints concerning creation. We shall now briefly discuss the 3 main interpretations accepted by various Bible-believing creationists.

(a) The Day-Age Theory

This theory proposes that the 6 days of creation represent periods of time, ages, not literal 24-hour days. This theory is designed to accommodate the geologic ages and is compatible with evolutionary thinking. It has been shown, however, to be an unworkable compromise, both Biblically and scientifically, and is riddled with numerous serious difficulties. For example, if the days are actually ages, how did the fruit trees created on the third day survive for ages before the sun was created on the fourth day? Similarly, this theory fails to accomodate the vital symbiotic inter-relationships among plants (third day), birds (fifth day), and insects (sixth day). For further information, see *Scientific Creationism,* by Henry M. Morris.

(b) The Gap Theory

This view has also been known as the *Ruin-Reconstruction Theory.* Proponents of this scheme maintain that the proper translation of Genesis 1:2 should actually be rendered: "Now the earth *became* formless and empty." The implication is that the original perfect creation came to a sudden and terrible cataclysmic ruin (usually associated with the fall of Lucifer). The earth is considered to be ancient with much of the geologic ages transpiring between Genesis 1:1 and 1:2 in an imaginary gap prior to the six days of *re-creation.* Like the Day-Age Theory, the Gap Theory endeavors to reconcile Biblical creationism to the evolutionary framework. It, too, is plagued with many scientific and Biblical contradictions and blunders and is not the preferred interpretation. For additional information, see *Scientific Creationism,* by Henry M. Morris.

(c) The Literal-Historical Theory

This interpretation regards the six days of creation as literal twenty-four hour days that followed in immediate succession. The earth is generally believed to be only a few thousand years old; the geologic ages and the concept of organic evolution are completely rejected. Proponents place great emphasis on the Genesis Flood when interpreting earth history and geologic features. This scriptural interpretation has been supported by many knowledgeable people including Martin Luther, John Calvin, Henry M. Morris, and the Creation Research Society.[80] Unlike the Day-Age Theory and the Gap Theory, this viewpoint is supported both Biblically and scientifically. It is therefore favored as the correct interpretation in this work.

Creationists maintain that uniformitarian principles simply cannot account for most of the major geologic features and formations. For instance, there is the vast Tibetan Plateau which consists of sedimentary deposits, which are thousands of feet thick, located presently at an elevation of three miles above sea level. The Karoo formation of Africa contains an estimated 800 billion vertebrate animals! The Herring fossil bed of California contains approximately one billion fish within a four-square mile area. The uniformitarian concept is equally incapable of explaining the Columbia Plateau in northwestern United States which is an incredible lava plateau several thousand feet thick covering an area of 200,000 square miles. Uniformitarianism also fails to offer a reasonable explanation for important geological concepts such as mountain building.[25]

Proponents of the literal-historical creation theory maintain that the worldwide Genesis Flood is responsible for most of the important geologic features and formations of the earth. It is believed that the Flood was accompanied by massive and violent earth movements, volcanic action, and dramatic changes in climate and topography. Creationists attribute modern-day topography to sudden and supernatural causation *after* the Flood (Psalm 104:6-9) which helps to explain how sedimentary strata (formed during the Flood) have come to be lifted thousands of feet above sea level in the mountainous regions of the earth. Thus, the Genesis Flood is viewed as an incredible geologic event, absolutely without parallel. It was a hydraulic episode that produced sedimentation and fossilization on a scale never approached before or since. Instead of being a record of gradual transformation, the geologic record is perceived as a chronicle of sudden and cataclysmic mass death and destruction. Properly understood, the geologic record is a fascinating testimony to the awesome power and judgment of God when, not so long ago, the world, which then was, perished beneath worldwide flood waters and their contained sediments.

The universal presence of fossils in sedimentary deposits is, indeed, conclusive proof of rapid burial and formation. Rapid burial and lithification are essential for the formation and preservation of fossils; otherwise, they would decay or be destroyed by scavengers. The fact that large scale fossilization is not occurring anywhere in the world today is a serious problem for uniformitarian geologists.

Sudden catastrophic deposition is indicated not only by the fossils contained in the sedimentary strata but also by the very strata themselves. According to Dr. Henry M. Morris, hydraulic analysis shows that, in most formations, the individual stratum has been formed within a few minutes time. Furthermore, there is evidence to suggest each subsequent stratum began to be deposited immediately after the preceeding one. This suggests that the geologic column was formed rapidly, not gradually over aeons of time. Thus, the geologic evidence supports a cataclysmic interpretation rather than a uniformitarian explanation.

To further substantiate this point, consider the fact that about ¾ of the earth's surface is covered with sediments or sedimentary rocks which were originally deposited under moving water. This is true even on the tops of many mountains. Clearly, the entire surface of the earth was once submerged by great and powerful waters. Only a global hydraulic cataclysm, such as the Genesis Flood, can account for the worldwide sedimentary rock deposits of sandstones, shales, conglomerates, limestones, dolomites, and evaporites. The abundance of gravels, conglomerates, and even boulders in many sedimentary formations is impressive evidence for extra-high intensity hydraulic activity. The common occurrence of cross-bedding, which signifies rapidly changing current directions, also serves to confirm this fact.

In the face of such overwhelming evidence for Biblical creationism and catastrophism, more and more evolutionary geologists are realizing that the present is *not* the key to the past. Many are now considering concepts such as local catastrophism to explain observed geologic features and formations. Evolutionists who believe major flooding offers the most reasonable explanation for the fossil record are known as *neo-catastrophists.* Still others, who have also recognized the necessity for a catastrophic explanation for many of the earth's features, have proposed some very imaginative catastrophic theories. Immanuel Velikovsky, for example, has proposed such a theory based upon a series of comet encounters. Velikovsky's ideas, however, are not only highly improbable but also lack a proper respect for the Scriptures, denying such fundamentals as miracles and the historicity of the Bible.

Although the evidence unquestionably points to catastrophism, there is no need to resort to neo-catastrophism or imaginative speculation when it comes to ascertaining the nature of the cataclysm. The Bible plainly provides the answer for us in the worldwide Genesis Flood.

In addition to the scientific and Biblical evidence for the Genesis Flood, it should also be mentioned that there are numerous ancient secular records (Babylonians, Sumerians, etc.) which tell of a great flood on the earth. There are also many flood traditions from tribes throughout the world. Thus, we find the Genesis Flood of the

Bible to be one of the best confirmed events in history.

8. Summary

It is now apparent that there is more than sufficient scientific and Biblical evidence to justify complete rejection of the geologic column and time-table. The geologic column, which is supposedly proof of evolution, is in itself actually founded upon the assumption of evolution; thus, it is blatantly entangled in a case of circular reasoning. Important recent findings directly contradict the fundamental precepts of the geologic column and time-table. Radiometric dating techniques are based on unreliable presuppositions and, therefore, offer no scientific validity to the supposed antiquity of the earth. Hence, we conclude that the widely accepted geologic column and time-table of earth history are essentially meaningless.

On the other hand, there is convincing data available to support the concept of a relatively young earth. Biblical creationism and catastrophism are much better suited to the facts at hand than evolutionary uniformitarianism. The creation model with its proper emphasis on the Genesis Flood and associated catastrophic geological events, is the only satisfactory explanation that can account for the observed complex of geologic structures, formations, and features in the earth today. Thus, the true evidence of geology does not testify to evolution at all but rather is a record of God's awesome power and righteous judgment on sin.

B. Paleontology

1. Introduction

The science of Paleontology supposedly offers the most impressive and convincing evidence for the theory of evolution. It is the branch of geology that deals with prehistoric life through the study of fossils. Evolutionists insist that it is the fossil record which provides actual documentation of organic evolution. But does the fossil record actually support evolution, or does it rather advance Biblical creationism and catastrophism? We intend to show that the concept of evolution is actually nothing more than a scientist's version of the old nursery tale about a frog being transformed into a prince. In this case, the magical transformation is not instantaneous, but requires the magical wand of geologic time (Figure 8). Ironically, as we shall endeavor to demonstrate, the fossil record does indeed testify strongly against evolution and favors Biblical creationism and catastrophism.

2. Sudden Appearance of Advanced Life Forms

The fossil record reveals an absence of life forms in the lower two-thirds of the earth's crust (the so-called pre-Cambrian period). Then suddenly, in abundant numbers of advanced forms, life appears. The oldest rocks in which indisputable fossils are found are those of the so-called Cambrian period. The Cambrian sedimentary deposits contain billions and billions of fossils of highly advanced and developed life forms. Everyone of the ma-

FIG. 8: A Scientist's Nursery Tale. Although evolutionists deny and ridicule the miraculous aspects of Biblical Creationism, they nevertheless insist that the evolutionary process, given enough time, produces the same miraculous results.

jor invertebrate forms of life have been found in Cambrian rocks. The complexity of these animals is so great that evolutionists estimate that they would require at least 1½ billion years to evolve. [25]

Obviously, if evolution is true, we should expect to find billions of evolutionary ancestors of the Cambrian life forms in pre-Cambrian rocks. However, as observed by the renowned biochemist, Dr. Duane T. Gish:

> "Not a single, indisputable, multicellular fossil has ever been found in pre-Cambrian rocks! Certainly it can be said without fear of contradiction that the evolutionary ancestors of the Cambrian fauna, if they ever existed, have never been found." [25]

In short, the fossil record reveals that life appeared abruptly in great diversity, complexity, and abundance without any ancestors from which to evolve! Clearly, this is not evidence for gradual organic evolution. This is supernatural creation.

3. Permanence of Kinds

Many fossils of plants and animals found in the supposed oldest of rocks, when compared with their living counterparts, are found to be essentially the same. In spite of presumed hundreds of millions of years of evolution, the present shellfish *Lingula,* starfish, cockroach, bacteria, and so on, are no different from their remote ancestors which allegedly lived, respectively, 500, 500, 250, and 600 million years ago! True to this very day are the words of Charles Darwin who conceded in his writings that:

> "Not one change of species into another is on record . . . we cannot prove that a single species has been changed." [19]

Much to the discontentment and embarrassment of

the evolutionists, the fossil record strongly supports the Biblical principle of reproduction *after its kind.* This fact strikes a shattering blow at the theory of evolution which requires life to be in a continual state of flux.

The Biblical statement, *after its kind,* not only is verified by the paleontological record but also is confirmed by modern scientific observation and experimentation. Practical breeders and geneticists have experimentally verified the great stability of kinds. Horizontal variation (sometimes refered to as *micro-evolution*), operating within limits specified by the DNA for the particular organism, is possible and has been used to develop many breeds. For example, there are over 200 varieties of dogs (Figure 9). It is possible to obtain 1,500 varieties of the Hawthorne plant by natural variation. Darwin's various finches may also be attributed to such horizontal variation. Mankind even has considerable potential for variation into various races. Consider the 4-foot pygmies of

FIG. 9: The Variety of Dogs. There are over 200 varieties of dogs ranging in size from the 4 pound Chihuahua to the 180 pound Great Dane. This does not imply evolution but rather demonstrates variation within a kind.

central Africa in comparison with the 9-foot Anakim (a race of giants descendent from Anak) of ancient Palestine (Numbers 13:28-33; Deuteronomy 2:10, 11, 21).

Vertical transformation (sometimes termed *macro-evolution*) of one kind of organism into an entirely new organism is, however, prohibited and does not occur. Dogs never change into horses, Hawthorne plants never become roses, and finches never become anything but other finches. Centuries of breeding experiments provide solid evidence which argues convincingly against the theory of organic evolution. Boundaries between kinds are very real and stubborn biological facts. When abnormal crosses are attempted, sterility is always the result:

Horse	+	Donkey	=	Sterile Mule
Zebra	+	Horse	=	Sterile Zebronkey
Lion	+	Tiger	=	Sterile Liger [39]

The fact that each of these hybrids are invariably sterile and do not reproduce is indeed strong evidence against evolution. It should be mentioned that the Biblical term *kind* bears no direct correlation to the arbitrary Linnaean taxonomic system although some feel that the *family* designation may be a reasonably close approximation. Gish defines the Biblical term *kind* as:

> "A generally interfertile group of organisms that possesses variant genes for a common set of traits but that does not interbreed with other groups of organisms under normal circumstances." [28]

One final consideration regarding permanence of kinds involves *Drosophilia,* the fruit-fly. The fruit-fly has been bred in the laboratory for over 1,000 generations, being subjected to continual radiation bombardment. One would suppose that after such given circumstances some kind of evolutionary development would arise. Yet, while it has managed to produce a great variety of mutational deformities, no new evolutionary life forms have been produced; it is still a fruit-fly.[53]

Once again, we find that the actual observed facts of science serve to confirm the Biblical record. Created organisms reproduce after their own kind and not after some other kind, with a limited amount of variation permitted within the permanently fixed kind (Genesis 1:11, 12, 21, 24, 25). This principle is also confirmed in the New Testament as we read in 1 Corinthians 15:38, 39:

> "But God giveth it a body as it hath pleased Him, and to every seed his own body. All flesh is not the same flesh: but there is one kind of flesh of men, another flesh of beasts, another of fishes, and another of birds."

4. Absence of Transitional Forms

We now come to perhaps the most serious of defects in the evolutionary theory—the complete absence of transitional forms. If life has always been in a continual stream of transmutation from one form to another, as evolutionists insist, then we should certainly expect to find as many fossils of the intermediate stages between different forms as of the distinct kinds themselves. Yet, no fossils have been found that can be considered transitional between the major groups or phyla! From the very beginning, these organisms were just as clearly and distinctly set apart from each other as they are today. Instead of finding a record of fine gradations preserved in the fossil record, we invariably find large gaps. This fact is absolutely fatal to the general theory of organic evolution. Even the great champion of evolution himself, Charles Darwin, acknowledged this fatal flaw:

> "As by this theory, innumerable transitional forms must have existed. Why do we not find

them imbedded in the crust of the earth? Why is all nature not in confusion instead of being as we see them, well-defined species? Geological research does not yield the infinitely many fine gradations between past and present species required by the theory; and this is the most obvious of the many objections which may be argued against it. The explanation lies, however, in the extreme imperfection of the geological record." [20]

While Darwin was honest enough to admit the reality and seriousness of the missing links, he nevertheless felt or hoped that this was only due to an incomplete fossil record. In time, he argued, these connecting links would be found and the critical gaps filled. This convenient excuse, however, no longer offers any refuge for evolutionists. As George Neville remarks:

> "There is no need to apologize any longer for the poverty of the fossil record. In some ways it has become almost unmanageably rich, and discovery is outpacing integration . . . The fossil record nevertheless continues to be composed mostly of gaps." [58]

Professor N. Heribert-Nilsson of Lund University, Sweden, has studied the subject of evolution for over 40 years and has commented on this problem of missing links:

> "It is not even possible to make a caricature of evolution out of paleobiological facts. The fossil material is now so complete that the lack of transitional series cannot be explained by the scarcity of the material. The deficiences are real, they will never be filled." [30]

Ironically, although it has been over 100 years since Darwin's time, we now have fewer examples of transitional forms than we did then. Instead of having more (as Darwin had hoped), we actually have less. This is because some of the old classic examples of evolution have been recently discarded due to new information and findings, and no new transitional forms have been found. Despite these insurmountable problems, the dauntless faith of the evolutionists persists. A. Lunn once wrote a humorous parody of such faith:

> "Faith is the substance of fossils hoped for,
> the evidence of links unseen." [67]

The well-developed and firmly established science of taxonomy ironically lends support to Biblical creationism. How so? Consider for a moment the basic proposition of the evolutionary model that life is in a continual state of flux, ever changing through infinitesimally small mutational changes. If this was true, then classification would be impossible. But the fact that living organisms are distinctly different and easily classified into separate categories is in complete harmony with the creation model.

According to the general theory of evolution, the basic progression of life leading to man was in the following manner:

1. Non-living matter
2. Protozoans
3. Metazoan invertebrates
4. Vertebrate fishes
5. Amphibians
6. Reptiles
7. Birds
8. Fur-bearing quadrupeds
9. Apes
10. Man

Now, as we have already established, if these things actually happened, it is perfectly logical and reasonable to expect that we should find vast numbers of transitional forms objectively preserved in the fossil record. This, however, is not the case and the supposed transitional forms between the major groups are missing in every case (Figure 10). Consider the following immense gaps:

(a) The imagined jump from dead matter to living protozoans is a transition of truly fanciful dimension, one of pure conjecture.

(b) There is a gigantic gap between one-celled micro-organisms and the high complexity and variety of the metazoan invertebrates.

(c) The evolutionary transition between invertebrates and vertebrates is completely missing. This is absolutely incredible since evolutionists propose 100 million years of developmental time between the two, which would have involved billions of transitional forms. Yet, not one such transitional form has ever been found.

(d) The evolutionary advance from fishes to amphibians is totally nonexistent. The evolution of fishes

FIG. 10: Missing Links. Wanted, dead or alive, are the numerous missing evolutionary links. Their complete absence deals a fatal blow to the general theory of organic evolution.

into amphibians supposedly took about 30 million years, and yet, no one has been able to produce even one *fish-ibian.* The coelacanth was once cited as an intermediate, but has been subsequently disqualified. Instead of being extinct for millions of years, the coelacanth was discovered to be still very much alive in 1938.

(e) There are no connecting links between amphibians and the altogether different reptiles. Seymouria has been offered as such a link, but it supposedly occurs in the geologic column some 20 million years after other reptiles had already appeared.[78]

(f) There are no transitional forms between reptiles and mammals.

(g) There are no connecting evolutionary links between reptiles and birds. Archaeopteryx was once highly acclaimed as such a link but has since been acknowledged by paleontologists to have been a true bird. [54]

(h) There are no intermediate or transitional forms leading up to man from an apelike ancestor. Fossil hominoids and hominids cited by evolutionists to demonstrate human evolution are actually fossils either of apes or men, or neither. There is no valid scientific evidence to suggest that they are fossils of animals intermediate between apes and men.[54]

In an attempt to explain the lack of transitional forms, some scientists have recently proposed the idea that evolution occurs via sudden large leaps rather than through gradual small modifications. This concept, known as *punctuated equilibrium,* has been advanced by paleontologists Gould and Eldridge (1977).[28] This concept has also been termed the "hopeful monster" mechanism by Goldschmidt who proposes that at one time a reptile laid an egg and a bird hatched from it! Creationists prefer to believe that these scientists are the ones who have laid the egg, maintaining that such ideas are pure speculation, completely devoid of any scientific evidence.

To summarize, the fossil record reveals a sudden appearance of highly diverse and complex forms with no evolutionary ancestors, demonstrates fixity of kinds, is devoid of the all-important transitional forms, and provides sufficient justification for rejection of the evolutionary theory. These facts are, however, in complete agreement with the expectations of the Biblical creation model. The rocks do indeed proclaim, "Creation!"

5. Fossilization

The fossil record presents yet another serious problem for evolutionary uniformitarianism: large-scale fossilization is not occuring anywhere in the world today! When a fish dies, it does not sink to the bottom and become a fossil. Instead, it either decomposes or is destroyed by scavengers. Similarly, there is hardly a trace left of the millions of buffalo carcasses, which were slaughtered all over the plains only 2 generations ago.

In sharp contrast to the virtual lack of fossilization transpiring today, there is an incredible amount of fossilization that occurred sometime in the past. The billions and billions of fossils we find preserved in the fossil record simply could not have been formed by processes observable in the world today as uniformitarian geologists have assumed and taught. Such preservation is very abnormal, the exception, and not the rule. Our global fossil record, therefore, attests to a strictly atypical world-wide, cataclysmic, hydraulic event. The fossil record of the geologic column is not a history of evolving life forms, but rather it is a great memorial to the sudden mass extermination of life from another age—the annihilation of the antediluvian world, around 2,346 B.C., by the Genesis Flood. [22] As the Scriptures record:

> ". . . every living substance was destroyed which was upon the face of the ground, both man, and cattle, and the creeping things, and the fowl of the heaven; and they were de-

stroyed from the earth: and Noah only remained alive, and they that were with him in the ark."

(Genesis 7:23)

Evolutionists insist that the progression of life forms found in the fossil record from simple to complex prove the evolutionary progression of life. However, it should be pointed out that the same general progression would be expected from the hydraulic sorting action of a worldwide cataclysmic flood. Creationists reason that the progression from marine invertebrates to fishes, to amphibians, to reptiles, to mammals, to man simply reflects increasing ability to escape death and destruction from the Genesis Flood. Naturally, exceptions would be expected to this general rule of hydraulic, ecologic, and physiologic sorting in such a chaotic event. While such exceptions would be expected and could be accomodated in the creation model, such reversals present a serious problem to evolutionists since they represent a reversal in the presumed evolutionary progression of life.

There are numerous spectacular examples of fossilization that corroborate the Genesis account of cataclysmic destruction. Animals are commonly found buried in an attitude of terror with heads arched back, mouths open, etc. We will now briefly consider a few such examples.

(a) Fossil Graveyards

There are caves, fissures, and mass burial sites throughout the world that are literally packed with masses of fossils; often times the fossils of these various animals come from widely separated and differing climatic zones, only to be thrown together in disorderly masses. Such phenomena can only be satisfactorily explained in terms of a worldwide aqueous cataclysm.

For example, consider Cumberland Cavern in Maryland that contains the remains of various animals from warm, damp, semi-tropical areas, arid regions, and from cold, northern zones. The Norfolk forest-beds in England contain the mixed remains of northern cold-climate animals, tropical warm-climate animals, and temperate zone plants. Other similiar deposits would include the Baltic amber deposits, the Agate Spring Quarry in Nebraska, the Rancho La Brea Pit in California, the Old Red Sandstone in Scotland, numerous rock fissure sites in England and France, and many others throughout Europe. [36]

Uniformitarian principles utterly fail to offer any explanation for such bizarre conditions. These tremendous fossil graveyards and mass burial sites can only be correctly understood in terms of Biblical catastrophism with its proper emphasis on the worldwide Genesis Flood. Clearly, these are the fossil remains of the antediluvian plants and animals that were destroyed, transported, and deposited by the powerful waters of the cataclysmic Genesis Flood.

(b) Mammoths

Further startling evidence of the fact that a great and sudden cataclysm once struck the earth is found in the millions of mammoths and other large animals that were killed instantly in the north polar regions (northern Siberia and Alaska). Many of these have been found preserved whole and undamaged (except for being dead, of course) with flesh and hair intact, and in some cases, either kneeling or standing upright with food on their tongues. The eyes and red blood cells were found to be extremely well preserved, and the separation of water in the cells was only partial, which speaks of extremely sudden and sustained freezing conditions (Figure 11).

All uniformitarian explanations fail dismally when attempting to interpret this phenomenon. Charles Lyell, 19th century champion of uniformitarianism, was aware of this problem and recognized the seriousness of the threat to his theory. He suggested that they were caught in a cold snap while swimming. This explanation, however, is obviously inadequate and simply does not fit the observed facts. Charles Darwin was also aware of the mammoths and confessed that he saw no solution to it. There is no uniformitarian solution to this problem; the evidence unquestionably requires a sudden catastrophic explanation.

Creationists propose a possible explanation for this amazing condition—the collapse of a vast antediluvian vapor canopy that surrounded and enveloped the pre-Flood World (Genesis 7:11). It is believed that such a canopy would have produced a worldwide greenhouse effect. The climate would have been mild throughout the earth with insignificant seasonal variations. No rainfall or rainbows would have existed but rather a mist would rise from the earth that would water the whole face of the ground (Genesis 2:5-6; 8:22; 9:13). Violent storms, such as we experience throughout the world today, would have also been precluded by this canopy (Genesis 2:5-6; Hebrews 11:7). Such a canopy would also help to explain why palm leaves, fruit trees, tropical marine crustaceans, coral reefs, and vast amounts of sub-

FIG. 11: A Frozen Mammoth. The existence of an estimated 5,000,000 frozen mammoths along the coastline of Northern Siberia and Alaska testifies of a sudden cataclysm which struck the earth in the recent past. There is no satisfactory uniformitarian explanation for this phenomenon.

tropical plant life are buried under the polar regions.[62] It is also believed that this vapor canopy may have served to filter out harmful radiation from space which is known to have an accelerating effect on mutations and the aging process. The existence of this canopy may well have been the key factor accounting for patriarchal longevity before the Flood (Genesis 5:5-27). After the Flood, the ages of the various patriarchs in the Bible exhibited a steady decline:

Noah	=	950 years
Salah	=	433 years
Peleg	=	239 years
Abraham	=	175 years
Moses	=	120 years
David	=	70 years
Present	=	70 years (Ps. 90:10)

Creationists maintain that only the collapse of such a tremendous vapor canopy can explain the statement found in Genesis 7:11-12, that:

> ". . . the windows of heaven were opened. And the rain was upon the earth forty days and forty nights."

This cannot refer to common rainfall for if all the water vapor and clouds in the present atmosphere were precipitated to the earth, the rainfall would amount to only a few inches and end after just a few short hours. Thus, this reference can only refer to the collapse of a gigantic antediluvian vapor canopy.

What does all of this have to do with all those poor frozen mammoths up in Siberia? Well, it is quite possible that the collapse of the vapor canopy may have been responsible for the sudden, almost instantaneous, freezing of the mammoths in the polar regions. Creationists theorize that temperatures of up to 150 degrees below zero could have descended upon them instantly during the early phases of the collapse of the vapor canopy. [73] While such a theory cannot be proven, it does have considerable scientific and Biblical support and is certainly better suited to the facts at hand than any uniformitarian explanation. For further detailed information regarding the earth's pre-Flood vapor canopy, see *The Waters Above,* by Joseph C. Dillow (1981).

(c) Extinctions

For many years evolutionists have been baffled by the fact that strong, well-established groups of animals such as dinosaurs and trilobites suddenly disappear from the fossil record. Although their sudden departure puzzles evolutionists, creationists simply attribute their misfortune either directly or indirectly to the Genesis Flood. Creationists theorize that the collapse of the vapor canopy resulted in a post-Flood climate which was dramatically different from the pre-Flood climate. Instead of continuing to enjoy the stable, warm, and mild climate which they were accustomed to, these animals found themselves thrust into a relatively unstable and more hostile enviroment characterized by cooler temperatures, severe storms, and bitter winter conditions. Certain animals simply could not make the adjustment to this new climate and eventually succumbed as a group to extinction. Cold-blooded animals, such as dinosaurs, whose body temperature is regulated by the temperature of the environment were probably especially affected in an adverse manner by the collapse of the vapor canopy and the subsequent change in climate.

(d) Petrified Logs

Petrified logs present a troublesome enigma for uniformitarian scientists because they are not being formed anywhere in the "uniformitarian world" today. They do, however, occur by the thousands in the fossil record having their fibers and cell structure perfectly preserved by replacement of silica.

All of the evidence concerning these logs points to a sudden and catastrophic event. None of these logs are found to be standing. All have had their branches stripped off and many appear to have had their bark still intact, indicating rapid burial before rotting could occur. It is apparent that the original forests were uprooted by some sort of hydraulic cataclysm of enormous power, which also transported and deposited them in their present locations, where they became petrified (under unique conditions). Creationists submit that the only reasonable and plausible explanation for such a phenomenon is the Genesis Flood.[73]

(e) Polystratic Trees

Polystratic trees are fossil trees which extend through several layers of strata, often 20 feet or more in length. There is no doubt but that this type of fossil was formed relatively quickly; otherwise it would have decomposed while waiting for strata to slowly accumulate around it. In some cases, these trees bridge a presumed evolutionary time span of millions of years. Obviously, the more reasonable interpretation calls for the simultaneous transportation and deposition of the trees and their surrounding sediments. Such logical conclusions cast serious doubts upon the common uniformitarian assumption that these sedimentary strata were laid down gradually over millions of years. Once again, the evidence supports the Biblical catastrophism of the Genesis Flood.[48, 78]

(f) Ephemeral Markings

Ephemeral markings such as ripple-marks, rain imprints, worm trails, and animal tracks are found in great abundance in the fossil record. This special type of transient fossil is originally formed as an evanescent marking on the surface of a recently deposited layer of sediment. These structures are very perishable and easily destroyed by normal weather conditions or by erosion and sedimentation. The formation and preservation of ephemeral markings is not observed today under normal uniformitarian conditions. Their preservation depends entirely upon abnormally rapid and complete burial associated with chemical lithification processes. The fact that these structures are found in great abundance throughout the world in the fossil record testifies to a sudden and worldwide cataclysmic sedimentary event—the Genesis Flood.

(g) Soft Parts

There are numerous cases of preservation of actual soft parts (tissues) in the fossil record. This is true even in what is presumed to be the most ancient of strata. Such fossils are commonly discovered together in large masses. Dr. N.D. Newell, paleontologist of the American Museum of Natural History, has recently commented on such a fossil deposit:

> "One of the most remarkable examples of preservation of organic tissues in antiseptic swamp waters is a 'fossil graveyard' in Eocene lignite deposits of the Geiseltal in central Germany . . . More than six thousand remains of vertebrate animals and a great number of insects, molluscs, and plants were found in the deposits. The compressed remains of soft tissues of many of these animals showed details of cellular structure and some of the specimens had undergone but little chemical modification . . ." 59

Such remarkable deposits could not have been formed by normal, slow, uniformitarian, *in situ* processes. A-typical transportation of the organisms and rapid burial by sediments are clearly indicated. The uniformitarian assumption that such deposits of soft tissues have remained untouched by decay or erosion for millions of years seems absolutely ludicrous. The only reasonable alternative for such a phenemenon is a worldwide hydraulic cataclysm of relatively recent occurrence which would have quickly destroyed, transported, deposited, and lithified these organisms in the engulfing sediments. The Genesis Flood stands alone as the only sensible explantation for the observed facts.

6. Summary

To summarize, geologists have maintained that paleontology offers the most important evidence to substantiate the theory of evolution. However, as we have shown, the actual observed facts of the fossil record argue strongly against evolution and in favor of Biblical creationism and catastrophism.

The sudden appearance of advanced and diverse life forms that lack evolutionary ancestors, the permanence of kinds throughout time, and the complete absence of transitional forms in the fossil record testify vigorously for Biblical creationism. Likewise, the evidences of fossilization such as wooly mammoths, petrified logs, polystratic trees, ephemeral markings, soft parts, fossil graveyards, and so on, all designate the workings of a sudden and powerful worldwide aqueous cataclysm of unmatched proportions—the Genesis Flood! Thus, the evidence supplied by the fossil record substantiates the Biblical account of creation and the Flood while at the same time it soundly refutes evolution and uniformitarianism.

C. Uniqueness of the Earth

The existence of a creation necessitates the reality of a Creator. Intricate design requires a careful and intelligent master designer; organization requires an Organizer. This is pure and simple logic.

Let us suppose when the astronauts landed on the moon that they found a highly sophisticated computer system sitting among the rocks. Would it be reasonable and logical for them to conclude that it just happened to "evolve" through the fortuitous collisions of meteorites with local terrestrial rock formations or through some other such accidental event? Or would it be more logical to conclude that it was carefully and intelligently designed by a creator? Common sense dictates the latter explanation, of course! What about the chance occurrence of a blender transforming itself into the Apollo 16 moon rocket (Figure 12)?! Is this preposterous, insensible, ridiculous, and absurd? No more so than saying that out of accidental combinations of molecules, which somehow came into existence, man was accidentally assembled.

Similarly, the evidence that the earth-sun system was designed by God far outweighs any possibility that it all just happened to come together by mere chance. We will now consider a few features of the earth-sun system which appear to be specifically and very carefully designed for the unique purpose of supporting life:

1. The earth is positioned at just the right distance from the sun so that we receive exactly the proper amount of heat to support life. The other planets of our solar

FIG.12: The Apollo 16 Moon Rocket. Lift-off of the highly sophisticated Apollo 16 Moon Rocket. (NASA)

system are either too close (too hot) to the sun or else too far (too cold) to sustain life.[64]

2. Any appreciable change in the rate of rotation of the earth would make life impossible. For example, if the earth were to rotate at 1/10th its present rate, all plant life would either be burned to a crisp during the day or frozen at night. [64]

3. Temperature variations are kept within reasonable limits due to the nearly circular orbit of the earth around the sun. [36]

4. Temperature extremes are further moderated by the water vapor and carbon dioxide in the atmosphere which produce a greenhouse effect. [36]

5. The moon revolves around the earth at a distance of about 240,000 miles causing harmless tides on the earth. If the moon was located 1/5th of this distance away, the continents would be completely submerged twice a day! [64]

6. The thickness of the earth's crust and the depth of the oceans appear to be carefully designed. Increases in thickness or depth of only a few feet would so drastically alter the absorption of free oxygen and carbon dioxide that plant and animal life could not exist. [64]

7. The earth's axis is tilted 23½ degrees from the perpendicular to the plane of its orbit. This tilting, combined with the earth's revolution around the sun, causes our seasons, which are absolutely essential for the raising of food supplies. [64]

8. The earth's atmosphere (ozone layer) serves as a protective shield from lethal solar ultraviolet radiation, which would otherwise destroy all life.

9. The earth's atmosphere also serves to protect the earth from approximately 20 million meteors that enter it each day at speeds of about 30 miles per second! Without this crucial protection the danger to life would be immense. [64]

10. The earth is the perfect physical size and mass to support life, affording a careful balance between gravitational forces (essential for holding water and an atmosphere) and atmospheric pressure. [64]

11. The two primary constituents of the earth's atmosphere are nitrogen (78%) and oxygen (20%). This delicate and critical ratio is essential to all life forms. [64]

12. The earth's magnetic field provides important protection from harmful cosmic radiation.

13. The earth is uniquely blessed with a bountiful supply of water that is the key substance of life due to its remarkable and essential physical properties.

Many other examples of this type could be cited that would also support the idea that the earth was created and carefully designed for a purpose. Such numerous perfect and complex combinations of interrelated con-

FIG. 13: The Earth. A spectacular view of our home, Earth, photographed during NASA's Apollo 17 Lunar Landing Mission. (NASA)

ditions and factors essential to delicate life forms unequivocally point to intelligent purposeful design. To believe that such an intricately planned and carefully balanced life support system is the result of mere chance is absolutely senseless. Surely, the *honest* and *objective* observer has no other recourse than to conclude that the earth-sun system has been carefully and intelligently designed by God for man. As it is written:

> "The heaven, even the heavens, are the Lord's: but the earth hath He given to the children of men." (Psalm 115:16)

2.

PHYSICS

A. Introduction

The geological evidences presented in the previous chapter have inflicted irreparable damage to the general theory of organic evolution. We now turn to the scientific discipline of physics to ponder additional factual evidence, which also serves to refute evolution and support Biblical creationism. These facts are taken from the first and second laws of thermodynamics. These two laws are proven scientific laws which have been tested repeatedly under all types of systems. No reputable scientist doubts their validity and full applicability.

B. The First Law of Thermodynamics

The first law of thermodynamics is known as the *Law of Energy Conservation.* It states that energy can be converted from one form into another, but it can neither be created nor destroyed. This law teaches conclusively that the universe did not create itself! There is absolutely nothing in the present economy of natural law that could possibly account for its own origin. This scientific fact is in direct contradiction with the basic concept of naturalistic, innovative evolution. The present structure of the universe is one of conservation, not innovation as required by the theory of evolution.

Although scientists cannot account for the origin of energy and matter or why the total energy is conserved, the Bible offers an explanation. God alone can truly *create.* Man can only re-fashion pre-existing materials. Since God has ceased from His creative works (Genesis 2:3), energy can no longer be created. The reason energy cannot be destroyed is because God is "upholding all things by the word of His power" (Hebrews 1:3). He preserves and keeps in store His creation (Nehemiah 9:6; 2 Peter 3:7).

C. The Second Law of Thermodynamics

After being stunned by the first law of thermodynamics, the theory of evolution is to receive its fatal blow from the second law of thermodynamics. The second law of thermodynamics is known as the *Law of Energy Decay.* It states that every system left to its own devices tends to move from order to disorder (Figure 14). In other words, the universe is proceeding in a downward,

FIG. 14: The Second Law of Thermodynamics. All processes of nature have a tendency toward decay and disintegration. This overall increase in disorder is known as the Second Law of Thermodynamics.

degenerating direction of decreasing organization. Material possessions deteriorate and all living organisms eventually return to dust, a state of complete disorder. Given enough time, all of the energy of the universe will become random low-level heat energy and the universe will have died what is commonly referred to as a *heat-death.*

A process that results in a more ordered and complex product, contrary to the second law of thermodynamics, might be possible but would necessarily be very limited, rare, and temporary in effect. But evolution requires billions of years of constant violations of the second law of thermodynamics to be considered even remotely feasible! Thus, we find that **the second law of thermodynamics renders the theory of evolution not only statistically highly improbable, but virtually impossible.** In the words of British astronomer, Arthur Eddington:

> ". . . if your theory is found to be against the second law of thermodynamics I can give you no hope; there is nothing for it but to collapse in deepest humiliation." [23]

The principle of increasing entropy (increasing disorder and randomness) from the second law of thermodynamics is interpreted by many creationists to be a direct result of the curse placed on creation due to the Fall of man (Genesis 3:17-19). Creationists also believe that the creation will ultimately be released from this bondage to decay and corruption (Romans 8:18-23).

The second law of thermodynamics constitutes a grave problem for evolutionists, and it is not surprising to find that they usually choose to ignore it. When pressed for an explanation, two arguments are usually given to circumvent this law of nature.

The first argument is that the second law does not apply to open systems such as the earth. The argument is that the sun supplies the earth with more than enough energy to offset the loss of energy due to entropy. Although this may at first seem to be a reasonable argument, it has two major flaws. First, as Dr. Henry M. Morris points out, it confuses *quantity* of energy with *conversion* of energy.[54] Naturally there is enough energy to fuel an imagined evolutionary process, but that is not the question. The question is *how* does the sun's energy sustain evolution. The mere availability of energy does not automatically insure the development of orderly structural growth. Some kind of directional program mechanism is required to transform energy into the energy required to produce increased organization. For instance, a pile of lumber, bricks, nails, and tools will not automatically evolve into a building apart from a directing code, despite the fact that it is an open system receiving more than enough energy from the sun to carry out the job. And remember, a complex building is impossibly primitive compared with even the simplest living cell. Second, there is no such thing as a closed system. Therefore, to argue that the second law is inapplicable to open systems such as the earth is meaningless since all other systems are also open.

The second argument used to reconcile the entropy principle with evolution is that the second law does not apply to living systems. The phenomena of life, admittedly, does appear to exhibit a remarkable contrast to the entropy principle. A seed, for example, develops into a tree and an embryo grows into an adult. However, as Dr. Henry M. Morris points out, the growth process is actually not a contradiction of the second law:

> "The growth process is really only an outworking of the marvelous structure of the germ cell, which has within itself the encoded 'information' necessary to assimilate incoming chemicals and gradually build upon itself a structure like that of the parent organism. It does not really constitute an increase of order, but rather an outward manifestation of the marvelous complexity of the genetic system

> and the environmental energies it is able to utilize." [48]

Thus, we find that life really is not increasing in complexity contrary to the second law of thermodynamics. Rather adult organisms are simply the unfolding, outward expression of the pre-existing order in the genes. The blueprints for the growth and development of the adult organism were already present in the genes of the parents. The origin of life from this pre-existing order in DNA does not present any difficulty for the creationist. The evolutionist, however, finds himself faced with an indomitable problem. How did life begin without the pre-existence of such intelligent order and design? This question must forever haunt the atheistic evolutionist.

It should also be noted that apparent *decreases* of entropy can only be produced at the expense of a still greater *increase* of entropy in the external environment. Thus, the entire system as a whole continues to run down as required by the second law of thermodynamics. Furthermore, such processes are only temporary and eventually succumb to death and disintegration. Life forms attempt to postpone the second law of thermodynamics, but entropy eventually wins out. After all, biological systems and processes are merely complex chemical and physical processes, and to these the laws of thermodynamics do certainly apply. Dr. Harold Blum, an evolutionary biochemist, has recognized this fact and writes:

> "No matter how carefully we examine the energetics of living systems we find no evidence of defeat of thermodynamic principles, but we do encounter a degree of complexity not witnessed in the non-living world." [9]

Thus, we find that the second law of thermodynamics completely negates the concept of organic evolution. The creation model, however, predicts that the second law of thermodynamics will be operative and is thus, once again, substantiated by the facts of science.

D. Summary

The two most reliable scientific laws, the first and second laws of thermodynamics, prove that conservation and deterioration are the processes that characterize and direct the physical universe. These facts are in direct contradiction with the expectations and requirements of the evolutionary framework which hopes for a universe which is getting better and better, progressing ever-upward. Thus, the evolutionary model of origins is scientifically indefensible. It is now seen as an antiquated theory which has finally crumbled beneath the ever-accumulating weight of evidence against it. At the same time, however, Biblical creationism does correlate with the evidence at hand. We conclude with a noteworthy quote from Dr. Henry M. Morris:

> " . . . the Second Law proves, as certainly as science can prove anything whatever, that the universe had a beginning. Similarly, the First Law shows that the universe could not have begun itself. The total quantity of energy in the universe is a constant, but the quantity of available energy is decreasing. Therefore, as we go backward in time, the available energy would have been progressively greater until, finally, we would reach the beginning point, where available energy equalled total energy. Time could go back no further than this. At this point both energy and time must have come into existence. Since energy could not create itself, the most scientific and logical conclusion to which we could possibly come is that: 'In the beginning, God created the heaven and the earth.'"[49]

3.

MATHEMATICS

The theory of evolution proposes that all of the highly complex structures and systems of the universe are due to the operation of purely natural and haphazard processes of nature. No external supernatural agent (i.e., God) is needed or desired by the proponents of this naturalistic viewpoint. The universe is perceived as being self-contained and self-evolving.

In direct contradiction to this philosophy, Biblical creationism maintains that the innumerable, highly complex systems and intricate structures of the universe offer exceptionally strong evidence of an omniscient Creator. It is the creationist's view that the astounding degree of complexity and order found throughout the universe could never be produced by mere chance but rather represent the handiwork of an Almighty God.

In this chapter we shall consider the likelihood of the chance evolution of life utilizing the basic principles of mathematical probability.

Probability is simply the likelihood of an event occurring. For example, the probability of getting hit by lightning is about 1 in 600,000 (fortunately). The probability of winning a lottery grand prize with a single ticket is about 1 in 5.2 million (unfortunately).

Evolutionists insist that highly complex systems consisting of numerous inter-relating components can arise through purely random and aimless processes. To their way of "thinking," if enough monkeys typed for long enough, eventually one of them would type a perfect unabridged dictionary (Figure 15). Of course, this idea is completely nonsensical, and a brief consideration of probability statistics will reveal the absurdity and naiveté of such a viewpoint.

To illustrate, consider the likelihood of spelling the word "evolution" by randomly selecting nine letters from the alphabet. The probability is 1 chance in (26)∧9 trials (∧ represents exponentiation). This is equivalent to 1 chance in five trillion, four hundred and twenty-nine billion, five hundred and three million, six hundred and eighty thousand! For such a modest request (accidentally spell a nine-letter word) these are rather bleak odds.

Consider a series of 20 cards which are numbered 1 through 20. If these cards are thoroughly shuffled and then laid out successively in a line, the chance of laying them

FIG 15: Monkey Business. Evolutionists believe that life originated and developed through a purely accidental and aimless natural process. The probability of life arising in such a manner is comparable to the probability of a monkey typing a perfect unabridged dictionary.

down in numerical order from 1 to 20 is 1 in 2,432,902,008,176,640,000! This huge number is known as *20 factorial* (20!) and can be calculated easily by multiplying together all the numbers from 1 to 20.

The probability of accidentally generating Genesis 1:1 — "In the beginning God created the heaven and the earth" — is 1 chance in (26)∧44 trials. This is equivalent to 1 chance in 1.81479392 x (10)∧62 trials. In other words, the chance of randomly producing the first verse of the Bible is 1 in 181,479,392,000,000,000,000,000,000,000,000,000, 000,000,000,000,000,000,000,000,000.

Obviously, as the number of components increases, the probability of getting the desired result decreases rapidly. For example, let us consider the chance development of a very simple system composed of only 200 integrated parts (simple compared with living systems). The probability of forming such an ordered system is 1 in 200 factorial, or 1 chance in approximately 1,000,000, 000,000,000. This colossal number can be written more simply as 10^{375}. Thus, there is only 1 chance out of 10^{375} of selecting the *proper* arrangement for a 200-part integrated system on the first trial. But what if we keep on trying different combinations over and over again? Won't we eventually achieve the desired result? Well, to begin with, there are only 10^{80} electrons in the universe.[53] Assuming this to be the maximum number of parts available to work with, only $1 \times 10^{80} / 2 \times 10^{2} = 5 \times 10^{77}$ groups of 200 parts each could be formed at any one given time. But we have to form 10^{375} such groups to be certain of getting the correct one. Assuming that none of the first trial groups work, let us continue trying over and over again at a generous rate of 1 billion (10^9) trials per second. Furthermore, to give the evolutionists every possible advantage, let us keep on trying for a period of 30 billion years (10^{18} seconds) since this is the presumed age of the universe. But even granting such liberal concessions, we find that the maximum number of trial combinations which could be attempted is still only $(5 \times 10^{77})(10^9)(10^{18}) = 5 \times 10^{104}$. This is far too short of the needed 10^{375} trial combinations required for success. And so, even after all this, the chance that 1 of these 5×10^{104} attempts would yield the desired result of a 200-part system is only 1 out of $1 \times 10^{375} / 5 \times 10^{104} = 2 \times 10^{270}$. Simply stated, the chance that a system composed of 200 integrated parts could develop by mere chance is for all practical purposes, non-existent.

And yet a 200-part system is ridiculously primitive compared with living systems. Modern research by NASA has demonstrated that the most basic type of protein molecule that could be classified *living* is composed of at least 400 linked amino acids. Each amino acid, in turn, is made up of a specific arrangement of four or five chemical elements, and each chemical element is itself a unique combination of protons, neutrons, and electrons! [54] Golay has demonstrated that the chance formation of even the simplest replicating protein molecule is 1 in 10^{450}. [27] Wysong has calculated the probability of forming the proteins and DNA for the smallest self-replicating entity to be 1 in $10^{167,626}$, even when granting astronomically generous amounts of time and reagents! [78] Can you imagine what the chance formation of a more complex structure or organ such as the cerebral cortex in the human brain would be? It contains over 10,000,000,000 cells each of which is carefully arranged according to a specific design, and each of which is fantastically complex in itself! [53]

For his defense, the evolutionist might move for a mistrial on the basis of being misunderstood. Rather than suggesting that our 200-part integrated system be suddenly organized all at once, he is proposing that it develops gradually through a step-by-step mutation/natural selection process. Unfortunately for the evolutionist, this argument only serves to make matters worse for his cause. The probability of organizing a 200-part system by this step-by-step process is 1 out of the number represented by the series 2! + 3! + 4! + . . . + 200! (the symbol "!" represents *factorial).* Obviously, this number is much larger than 200 factorial and the chance of our 200-part system developing by this step-by-step mechanism is far less than its chance of developing all at once which was, for all practical purposes, a zero probability.

Someone has stated that the probability of life arising by mere chance is comparable to the probability of a monkey typing a perfect unabridged dictionary. If we concern ourselves with letters only (disregard punctuation, accent marks, numbers, spaces, capitalization, etc.) we find that we need to accidentally select about 35 million letters in the correct sequence.

Thus, we find that it is mathematically impossible for even the most elementary form of life to have arisen by mere chance. Life is no accident. It is not even something which brilliant scientists can synthesize. The bewildering complexity of even the most basic organic molecules completely rules out the chance of life originating apart from super-intelligent design and planning. The most logical and reasonable conclusion which can be reached based on mathematical analysis is that complex, ordered systems, which so characterize the world in which we live, never happened by mere chance but are the handiwork of our Creator, Almighty God.

> "Great is our Lord, and of great power: His understanding is infinite."
>
> (Psalm 147:5)

4.

BIOLOGY

A. Introduction

The world in which we live is truly an amazing and fascinating place. It houses a cheetah cat which can run 70 mph; insects which sleep for 17 years; Weddell seals that can remain under water for 45 minutes, diving to depths of 1500 feet; eight-armed, ink-shooting octopuses that can eat their own arms and grow new ones; archer-fish that can shoot water 15 feet into the air and hit a bug; peregrine hawks that can swoop down on their prey at 150 mph; and so on. The list of remarkable organisms and their amazing abilities is practically endless. The earth does indeed contain a myriad of highly complicated and intricate "adaptions," which completely defy evolutionary explanation. Extraordinary ecological relationships between various groups of organisms and their respective environments as well as with other organisms can only be satisfactorily explained in terms of intelligent forethought and planning. Nature is teeming with innumerable examples of ingenious design and purposiveness. The idea of purpose and design in the living world is known as *teleology.*

Creationists view with a sense of awe the marvelous fit of organisms to their environments and their incredible inter-relationships. Such complex and refined "adaptations," they reason, could never result from an aimless, purposeless, step-by-step process such as naturalistic evolution. The widespread beauty, intricacy, and perfection of life forms found throughout the earth eloquently testify to the reality of a super-intelligent Creator and Designer.

B. Extraordinary Design in Nature

George Gallup, originator of the well-known opinion polls, once remarked that he could prove God's existence statistically:

> "Take the human body alone—the chance that all the functions of the individual could just happen, is a statistical monstrosity." [63]

Although neither time nor space permit us to explore all the incredible mysteries of the human body, we will nevertheless begin our study of extraordinary design by reflecting briefly upon one small member of this bewildering complex—the eye.

1. The Eye

Evolutionists are hard-pressed to explain the step-by-step chance evolution of the eye which is characterized by a staggering complexity (Figure 16). Furnished with automatic aiming, automatic focusing, and automatic aperature adjustment, the human eye can function from almost complete darkness to bright sunlight, see an object the diameter of a fine hair, and make about 100,000 separate motions in an average day, faithfully affording us a continuous series of color stereoscopic pictures. All of this is performed usually without complaint, and then while we sleep, it carries on its own maintenance work. [77]

The human eye is so complex and sophisticated that scientists still do not fully understand how it functions. Considering the absolutely amazing, highly sophisticated

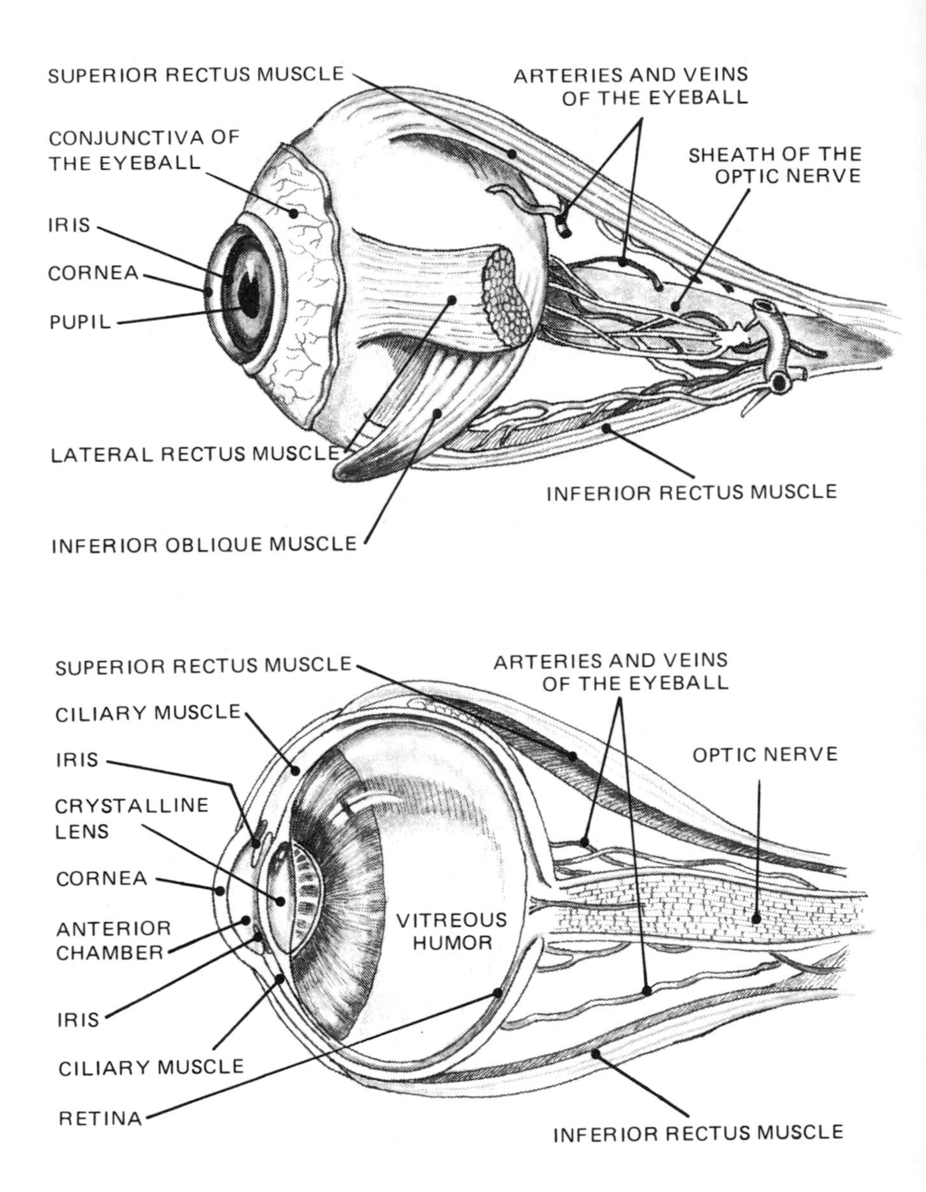

FIG. 16: The Eye. The eye is one of many marvelous examples of complex structures and organs which demonstrate incontrovertable evidence for design and purpose throughout creation.

synchronization of complex structures and mechanisms which work together to produce human vision, it is difficult to understand how evolutionists can honestly believe that the eye came about through the step-by-step, trial and error evolutionary process. This is especially true when we realize that the eye would be useless unless fully developed. It either functions as an integrated whole or not at all. Thus, the piecemeal evolution of the eye is completely outlandish and unreasonable.

Charles Darwin acknowledged the utter inadequacy of the evolutionary theory when attempting to account for a structure such as the eye:

> "To suppose that the eye, with all its inimitable contrivances for adjusting the focus to different distances, for admitting different amounts of light, and for the correction of spherical and chromatic aberation, could have been formed by natural selection, seems, I freely confess, absurd in the highest possible degree . . . The belief that an organ as perfect as the eye could have formed by natural selection is more than enough to stagger anyone." [67]

An incomprehensible constellation of favorable, integrated, and synchronized mutations would have to occur to produce an organ such as the eye. Granting evolutionists' generous concessions, Wysong nevertheless computes the probability for the chance formation of an eye at 1 in 10^{266}! [78] In light of these scientific facts, today's evolutionist would do well to abandon his dogmatic attitude and follow the honest example of Charles Darwin who conceded this serious flaw in the evolutionary theory.

The evolutionist's problems are further complicated by the fact that the evolutionary theory calls for the chance development of the eye several times, not just once. Frank Salisbury comments on this dubious prospect:

> "My last doubt concerns so-called parallel evolution . . . Even something as complex as the eye has appeared several times; for example, in the squid, the vertebrates, and the arthropods. It's bad enough accounting for the origin of such things once, but the thought of producing them several times makes my head swim." [66]

The inescapable conclusion is now quite evident. The eye did not just happen to develop at all but rather was created in the beginning by God in its complete and magnificent form. In the wise words of Sturmius, "examination of the eye is a cure for atheism." [60]

2. The Sea Slug

One of the most intriguing mysteries among marine creatures is found in the truly remarkable sea slug. The sea slug lives along the sea coast within the tidal zone where it feeds primarily on sea anemones. Sea anemones are not exactly the most inviting of dinners as they are equipped with thousands of small stinging cells on their tentacles which explode at the slightest touch, plunging poisoned harpoons into intruders. The speared intruder is paralyzed and drawn into the anemone's stomach to be digested.

Although this is an impressive defense system, the remarkable sea slug is able to eat sea anemones without being stung, exploding the stinging cells, or digesting them. One of the most fascinating mysteries in nature is what the sea slug does with the poor anemone's stinging cells. The undigested stinging cells are swept along through ciliated tubes which are connected to the stomach and end in pouches. The stinging cells are arranged and stored in these pouches to be used for the sea slug's defense! And so, whenever the sea slug is attacked, it defends itself using the stinging cells which the ill-fated anemone manufactured for its own protection.

The highly complicated series of modifications that would have had to occur to produce this incredible relationship completely defies evolutionary explanation. First of all, in order to prevent the stinging cells from exploding, the sea slug would have to evolve some sort of chemical means to temporarily neutralize them. The sea slug would also have to evolve a new digestive system, which would digest the tissues of the anemone but not the stinging cells. The sea slug would also have to cleverly evolve the sophisticated ciliated tubes and pouches as well as a highly complex mechanism for arranging, storing, and maintaining the stinging cells. Finally, and contrary to evolutionary expectations, the anemone would have to endorse the sea slugs plans by refraining from evolving countermeasures. [36]

Obviously, there is no satisfactory evolutionary explanation for the existence of such extraordinary adaptive design. The only reasonable solution to this fascinating relationship is offered by Biblical creationism. These organisms were specifically created and carefully designed by their Creator to fit into their respective ecological niches.

3. Gardening Ants

Another fascinating relationship, which has been observed in nature, concerns the Bull's Horn Acacia tree of Central and South America. This tree is furnished with large hollow thorns that are inhabited by a species of ferocious stinging ants. Small bumps on the tree also supply food to the ants. Consequently, the ants get food and shelter from the tree. The tree, for its part of the bargain, receives complete protection from all animal predators and plant competitors. The ants viciously attack any and all intruders. But the truly remarkable aspect of this symbiotic relationship is the fact that these ants are gardeners! They make regular raids in all directions from their home tree, nipping off every green shoot that dares to show its head near their tree. As a result, this particular tree always has plenty of sunlight and space which is a rarity in the tropical jungle where the competition for such things is intense. Experiments have shown that when all of the ants are removed from one of these trees, the tree dies within 2 to 15 months. [36]

Evolutionists casually label this type of symbiotic relationship as an example of *co-adaptation.* Of course, they never attempt to explain just how such an intricate relationship might have developed through the evolutionary process. Hence, the evolutionary viewpoint continues to be a matter of faith or presumption, not science. But how much more reasonable and logical is the faith of the creationist who perceives that these wonderful relationships are the purposeful and intelligently designed handiwork of an omniscient Creator.

4. Cleaning Symbiosis

An amazing relationship found in nature, which ridicules evolutionary thinking, is that of cleaning symbiosis. Fish, for example, roam about feeding on smaller fish and shrimp only to find that their mouths have become littered with debris and parasites. The solution to this problem for several types of fish is a visit to the local cleaning station.

At the cleaning station, the large fish opens its mouth and gill chambers, baring vicious-looking teeth, and in swim the undaunted little cleaner fish and shrimp to do their jobs. After their chore is completed, they swim back out of the larger fish's mouth unharmed, and the big fish swims away (Figure 17).

It is obvious that all parties involved benefit from this relationship, but this does not explain the origin and development of this special relationship. Survival value can only be used as an argument after a relationship has been established. The picture is further complicated for the evolutionist by the fact that several species of predatory fish, cleaner fish, and shrimp are involved in this operation.

Creationists maintain that this type of relationship could never have resulted from a mere chance, trial and error evolutionary process. Animal instincts for self-preservation would surely override any such unnatural suicidal tendency. Also, the temptation to get an easy free meal or to react to the irritating cleaners would tend to discourage the development of such a relationship. This type of arrangement can only reflect special creation.

FIG. 17: Cleaning Symbiosis. Cleaning symbiosis is one of many remarkable ecological interrelationships found in nature which challenges evolution, plainly testifying of special creation.

It should also be added that cleaning symbiosis is by no means limited to fish alone. Amazingly enough, there is a bird (the Egyptian plover) that is willing to walk right into the mouth of the Nile crocodile to clean out parasites. He, too, leaves completely unharmed. [60] Oh, the wonders of God's creation!

5. Beetle Warfare

Did you ever notice how sometimes big surprises can come in little packages? Well, such is the case of the surprising little bombardier beetle. The bombardier beetle is a small insect that is armed with a shockingly impressive defense system. Whenever threatened by an enemy attack, this spirited little beetle blasts irritating and odious gases, which are at 212° F, out from two tail pipes right into the unfortunate face of the would-be aggressor.

Dr. Hermann Schildknecht, a German chemist, studied the bombardier beetle to find out how he accomplishes this impressive chemical feat. He learned that the beetle makes his explosive by mixing together two very dangerous chemicals (hydroquinone and hydrogen peroxide). In addition to these two chemicals, this clever little beetle adds another type of chemical known as an inhibitor. The inhibitor prevents the chemicals from blowing up and enables the beetle to store the chemicals indefinitely.

Whenever our beetle friend is approached by a predator, such as a frog, he squirts the stored chemicals into the two combustion tubes, and at precisely the right moment he adds another chemical (an anti-inhibitor). This knocks out the inhibitor, and a violent explosion occurs right in the face of the poor attacker.

Could such a marvelous and complex mechanism have evolved piecemeal over millions of years? The evolutionist is forced to respond with a somewhat sheepish "yes," but a brief consideration of this opinion will reveal its preposterous nature.

According to evolutionary "thinking" there must have been thousands of generations of beetles improperly mixing these hazardous chemicals in fatal evolutionary experiments, blowing themselves to pieces. Eventually, we are assured, they arrived at the magic formula, but what about the development of the inhibitor? There is no need to evolve an inhibitor unless you already have the two chemicals you are trying to inhibit. On the other hand, if you already have the two chemicals without the inhibitor, it is already too late, for you have just blown yourself up. Obviously, such an arrangement would never arise apart from intelligent foresight and planning. Nevertheless, let us assume that our little beetle friend somehow managed to simultaneously develop the two chemicals along with the all-important inhibitor. The resultant solution would offer no benefit at all to the beetle, for it would just sit there as a harmless concoction. To be of any value to the beetle, the anti-inhibitor must be added to the solution. So, once again, for thousands of generations we are supposed to believe that these poor beetles mixed and stored these chemicals for no particualr reason or advantage; until finally, the anti-inhibitor was perfected. Now he is really getting somewhere! With the anti-inhibitor developed he can now blow himself to pieces, frustrating the efforts of the hungry predator who wants to eat him. Ah yes, he still needs to evolve the two combustion tubes, and a precision communications and timing network to control and adjust the critical direction and timing of the explosion. So, here we go again; for thousands of generations these carefree little beetles went around celebrating the 4th of July by blowing themselves to pieces until finally they mastered their new found powers.

But what would be the motivation for such disastrous, trial and error, piecemeal evolution? Everything in evolution is supposed to make perfect sense and have a logical purpose, or else it would never develop. But such a process does not make any sense at all, and to propose that the entire defense system evolved all at once is astronomically improbable, if not impossible. Yet, nature abounds with countless such examples of perfect coordination. Thus, we can only conclude that the surprising little bombardier beetle is a strong witness for special creation, for there is no other rational explanation for such a wonder. [24]

The water beetle is also equipped with an impressive—although different—defense mechanism. He manages to escape his enemies by secreting a detergent substance from a gland. Ejecting the detergent accomplishes two things. Firstly, it serves to propel the beetle forward quickly so that he is out of the immediate danger. Secondly, the detergent causes the surface tension of the water to break down, and the pursuing insect sinks into the water. [78] How true are the words of the Psalmist who wrote:

> "O Lord, how manifold are thy works! In wisdom hast thou made them all: the earth is full of thy riches."
>
> (Psalm 104:24)

6. Migratory Instincts

Migrating birds are capable of performing truly remarkable navigational feats. The lesser white-throated warbler is one such bird. It summers in Germany but winters in Africa. As the summer draws to a close and the young birds are independent, the parent birds take off for Africa, leaving their young behind. The new generation takes off several weeks later and flies instinctively across thousands of miles of unfamiliar land and sea to rejoin their parents. How do they manage to navigate with such precision across such distances especially since they have never been there before? Experiments have shown that within the brains of these birds is the inherited knowledge of how to tell latitude, longitude, and direction by the stars, plus a calendar, a clock, and all the necessary navigational data. All of this highly sophisticated equipment enables them to fly unguided to the exact location of their parents. [36]

Many other species of birds and other animals perform similar migratory feats. The golden plover travels some 8,000 miles south from the Hudson Bay region, crossing about 2,000 miles over the sea from Nova Scotia to the Caribbean countries, and winters in Argentina. It returns by way of Central America and the Mississippi Valley. The barn swallow migrates a distance of 9,000 miles from northern Canada to Argentina. The arctic tern migrates some 14,000 miles each year traveling from pole to pole and back. Whales, fur seals, bats, salmon, turtles, eels, lemmings, and various other animals also migrate. [62]

The causes of migrations and the incredible sense of direction shown by these animals presents the evolutionist with one of the most baffling problems of science. Evolutionists are indeed hard pressed to explain how these remarkable abilities evolved piecemeal through mere chance processes apart from any directing intelligence. The piecemeal development of such an instinct seems highly improbable because **migratory instincts are useless unless perfect.** Obviously, it is of no benefit to be able to navigate perfectly across only half of an ocean.

The perfection of migratory instinct, its intricacy, and its vital role in the widespread preservation of thousands of animals, simply cannot be explained in terms of gradual piecemeal evolution. The only logical conclusion, which can be reached to account for this remarkable phenomenon, is that these animals were carefully created and designed with these impressive abilities enabling them to lead enjoyable and successful lives generation after generation.

7. Insect Flight

Insects are the only invertebrates that possess the capability of flight, a fascinating phenomenon, which enables them to exist in a variety of environmental situations. They are more diverse than other invertebrates, having about one million species described.

The wing of an insect is a superbly designed flying tool that is capable of a very strong sculling action. Wing movement in insects is very complex, consisting of elevation and depression, fore and aft movement, pronation and supination, and changes in shape by folding and buckling. Many insects can hover or even fly backwards.

Some can even fly sideways or rotate about the head or tail by utilizing unequal wing movement.

Some insects, such as bees, wasps, and flies, must combine excellent flying skills with a small wing area. The honey-bee, for example, could not function in its hive with large wings. But the reduced wing area is compensated for by a very rapid wingbeat. Such frequencies range from 55 per second for some beetles, to over 200 per second for the honey-bee. The midge has an inconceivable wingbeat of 1,046 per second! [14]

Insect flight is truly an engineering wonder, which displays God's glory, power, and wisdom. As the Scriptures testify, the invisible things of God are revealed by the things of His creation so that man is without excuse if he rejects the truth (Romans 1:20).

8. Summary

Extraordinary design and adaptation are found throughout nature in such abundance, perfection, and complexity as to utterly defy evolutionary explanation. Most of these structures and abilities simply could not have evolved piecemeal because they are useful only when perfect and complete. Indeed, all of nature proclaims, "Behold, the Master Creator!" As Lord Kelvin has written:

> "Overwhelmingly strong proofs of intelligent and benevolent design lie around us . . . the atheistic idea is so non-sensical that I cannot put it into words." [35]

C. Visual Beauty

All of nature abounds with magnificent visual beauty. Many organisms exhibit beautiful coloration patterns and architectural designs. Where did all of this visual beauty come from, and why did it develop? Surely, the initial life form in the evolutionist's scheme did not exhibit such beauty. Creationists maintain that the origin, development, perfection, and widespread presence of visual beauty in the world of life completely defies evolutionary explanation. Evolutionists are especially hard pressed to account for the numerous instances in which beauty is hidden and unnecessary. Much to the consternation of the evolutionist, visual beauty is often quite useless except for the aesthetic gratification of man and God.

Many structures are beautifully colored despite the fact that they are rarely or never seen. For example, the abyssal fish, *Rhodicthys,* is of a bright red color. Yet, it lives in total darkness, 1½ miles below the surface of the ocean. Likewise, the deep-sea *Neoscopelus macrolepidotus* is vividly colored with azure blue, bright red, silver spots, and black circles! Even the eggs of some of the deep-sea creatures are brilliantly colored. [67]

Useless or hidden visual beauty is indeed an evolutionist's nightmare. To his way of "thinking," everything that evolves must have a practical purpose, or else it would never evolve. Evolution is supposed to be a responsible process. Why then the splendors of the abyssal fish, the beauty found inside some shells, the dazzling colors inside the mouths of nestlings, and so on? And remember, we have only considered color, saying nothing of the magnificent and limitless architectural designs found in nature. Also, if visual beauty is a natural consequence of evolutionary progression, why is it that the *lower* forms of life display greater visual beauty than the *higher* forms, such as man?

Evolutionary thinking fails dismally in attempting to explain visual beauty. The evolutionist is totally baffled by the world around him which is so characterized by perfect widespread beauty that often serves no practical function, except for aesthetics. The only logical and reasonable conclusion that agrees with the observed facts is that such beauty is the creation of the greatest Designer and Aesthete of all. As the Psalmist records:

> "The heavens declare the glory of God; and the firmament showeth His handiwork."
> (Psalm 19:1)

D. Mimicry

Mimicry is a fascinating phenomenon of nature in which one type of organism imitates or "mimics" another type. Most examples of mimicry are found among insects, although other animals and even some plants exhibit this capability. The extraordinary perfection, variety, and versatility of mimicry found in nature completely scorns the evolutionary philosophy.

The perfection of mimicry among insects is so great that it can successfully deceive a skillful naturalist who is watching for that very thing. In fact, it is often so perfect that it can bamboozle other insects being mimicked to the extent that the mimic can live among his enemies undetected! For example, spiders can disguise themselves as ants. While this may not sound particularly impressive at first, it is actually quite clever because spiders have eight legs whereas ants only have six legs and two antennae. To fool the ants, the spider holds his front pair of legs over his forehead and wiggles them like antennae. To further authenticate this deception, the spider also imitates the jerky gait and feeding movements of the ants. Some spiders have even been observed carrying the skeletons of ants over their bodies to disguise themselves.

The caterpillar of the Lobster moth of Britain is another fascinating example of mimicry. It has modified its legs to hang down like the scales surrounding the buds of the beech tree. Luckily (or so the evolutionist imagines) they are the proper number, length, color, and shape for this very purpose! When this larva is attacked, it lowers flaps on its sides which uncover black "wounds" which trick the attacker into thinking that it has already fallen victim to another parasite. Disappointed, the enemy departs. Such precise, intricate, and carefully reasoned mimicry leaves the evolutionist completely speechless and dumbfounded. [67]

Some mimics fool their predators by resembling stinging or bad tasting models, which the predators naturally avoid. For example, many species of butterflies imitate monarchs or other unpalatable butterflies, or moths. Some species of flies mimic bees or hornets. Equally bewildering to the evolutionist is the fact that sometimes only one sex of a species will mimic while the other will not.

Sometimes the predator rather than the prey becomes the mimic. One species of desert lizard entices insects to their death using the corner of its mouth, which when opened resembles a small desert flower. Similarly, the angler fish dangles a worm-like structure in front of itself to tempt other fish within reach of its hungry mouth. Certain female fireflies mimic the flashes of females of other species, and when the excited would-be suitor arrives, she eats him. The cuckoo of Europe and the cowbird of the United States both lay their eggs in the nests of other birds and successfully manage to have their young raised by the unsuspecting foster parents. Incidentally, the ungrateful foster nestlings eliminate all of the legitimate members of the family by pushing them out of the nest. [18]

Countless other examples of ingenious mimicry could be mentioned which also disallow any sort of evolutionary explanation. The abundance, variety, perfection, and resourcefulness of mimicry observed in nature almost defies comprehension or description. Evolutionary thinking is completely incapable of explaining the origin, development, or perfection of such mimicry especially when it is accompanied by the concomitant imitation of shape, color, habit, and so on. Without question, mimicry is indeed one of the most persuasive arguments against evolution and for special Creation.

E. Convergence

It is commonly known that similar organs, structures, colors, habits, and so on, exist in unrelated types of organisms. Evolutionists attribute this phenomenon to the effects of similar environmental demands on different organisms resulting in similar structures, habits, or whatever. This concept is known as *convergence.*

Creationists point out that it is difficult enough to account for the chance evolutionary development of a highly complex structure or organ just once. To suppose that such precise structures have evolved independently from widely different ancestral organisms, each time starting from a completely dissimilar structural beginning, is sheer fantasy.

There are numerous examples of convergence found in nature. For example, consider the case of the highly sophisticated sonar systems that are found in both the bat and porpoise. The extremely complex wing mechanism has supposedly evolved independently 4 seperate times (insects, bats, flying reptiles, and birds). To make matters even worse for the evolutionist, there are no fossil transitional forms that demonstrate the ability to fly as having evolved.

Willey has observed a striking example of convergence. He writes:

> "The most remarkable histological resemblance is manifested between the lateral sense-organs of the Capitellidae and the lateral sense-organs of Vertebrates (different phyla!). In both cases, the essential organs consist of small, solid, roundish, epidermal buds, from which fine, stiff sense-hairs project freely into the surrounding medium; and the resemblance is further enhanced by their segmental arrange-

> ment. The correspondence could hardly be greater, the convergence could hardly be closer, the homology could not be more remote than infinity." [76]

Such remarkable cases of convergence should shake even the ardent faith of the evolutionist. Instances of double convergence should shatter it completely. An example of double convergence is found in the coiled prehensile tail and independently rolling eyes of the sea horse and chameleons!

The piecemeal chance evolution of any complex mechanism or structure is highly improbable, if not impossible. The evolutionary development of numerous highly intricate structures, habits, and physiologies that are similar or identical, despite the fact that they arose from widely different anatomical origins, is completely inconceivable, an absolute impossibility. Cases of double convergence reveal the preposterous nature of such an unfounded concept. Clearly, evolutionary convergence must be abandoned and rejected as a complete absurdity.

Biblical creationism offers the only reasonable alternative. Creationists believe that the explanation lies in the fact that God, in some instances, chose to utilize the same or closely similar structures by various means in different organisms. Thus, convergence not only ridicules the evolutionary model, but it also serves as a subtle testimony to the unlimited capabilities and unparalleled virtuosity of our Creator:

> "Ah Lord God! Behold, thou hast made the heaven and the earth by thy great power and outstretched arm, and there is nothing too hard for thee."
>
> (Jeremiah 32:17)

F. Evolutionary Mechanisms

The most serious problem facing evolutionists is the dire need to find a plausible mechanism that might account for the evolution of life forms. Although several processes have been suggested over the years, none provide a satisfactory mechanism for organic evolution. The four primary mechanisms, that have been advanced to explain how evolution might occur, will now be briefly considered.

1. Lamarckism

Also known as the *theory of inheritance of acquired characteristics,* this theory was proposed by Jean Baptiste de Lamarck (1744-1829). This was the first systematized theory of organic evolution. The fundamental premise of this theory was that an organ that is constantly used will become more developed, while one that is not used will atrophy. These so-called *acquired characteristics* could then be inherited by succeeding generations until finally a new species would evolve.

Lamarck believed that the giraffe's long neck was developed in this very manner. As the giraffes fed on tree leaves, the constant stretching and reaching for higher leaves would tend to develop a slightly elongated neck. The offspring in turn, inherited this trait, improved upon it themselves, passed it on to their offspring, and so on.

It is now known, however, that change can only be transmitted to offspring through alterations in genes and their contained DNA. Discovery of this fundamental error of Lamarckism resulted in its ultimate rejection. By the 1930's, the theory of inheritance of acquired characteristics was fittingly discarded by the scientific community.

2. Darwinism

Charles Darwin formulated the mechanism, which we now refer to as *Darwinism,* in 1859. Also known as the *theory of natural selection,* this theory centers around the concept of survival of the fittest. Darwin observed that organisms are engaged in a constant struggle for survival, competing for food, water, shelter, and so on. He reasoned that since variation exists among organisms, those possessing more advantageous qualities will compete better and produce more offspring. Darwin would have interpreted the giraffe's long neck in terms of competition for survival. Giraffes with longer necks would compete for food more successfully and, therefore, produce more offspring like themselves. This process, however, is no longer accepted by scientists as the sole mechanism responsible for the supposed evolution of life forms. Natural selection explains *survival of the fittest* but it does not explain *arrival of the fittest.*

It is also worth noting that in Darwin's two works, *The Origin of Species* and *The Descent of Man,* the phrase, "we may suppose," or some similar clause, occurs over 800 times.

3. The Mutation Theory

This theory was proposed by Hugo deVries in 1901 and was based on the pioneering work of Gregor Mendel on genetics in 1866. Mendel's classic experiments with flowered pea plants exposed a fundamental fallacy in Darwin's theory. Mendel discovered that when he interbred the second generation red-flowered plants, obtained as the offspring of his original cross, between red and white parent plants, he got white as well as red flowers. Darwin's theory postulated that the white characteristic was a new evolutionary development, one which the parents did not possess. Mendel, however, demonstrated conclusively that this characteristic was not novel at all. It was present all the time in the parent's generation as a recessive characteristic temporarily hidden by a more dominant gene.

Obviously, if Darwinism was to survive, it would have to face up to these new facts and adjust accordingly. Conveniently, it was proposed that genes could somehow change into completely new forms via the process of *mutation,* hence, the *mutation theory.*[3] According to this theory, new species are a result of favorable mutations (chance DNA alterations). Modern scientists, however, also reject this process as the sole mechanism for organic evolution.

4. Neo-Darwinism

Modern evolutionary theories are modified and updated versions of Darwinian thinking, hence the term *Neo-Darwinism.* Modern scientists, who propose this type of viewpoint, postulate that the combined effects of natural selection (Darwinism), mutations (The Mutation Theory), and geologic time could account for organic evolution. Neo-Darwinists believe that mutations supply the needed variants from which nature can preferentially select over aeons of time. They concede that neither mutations nor natural selection alone could account for the supposed evolutionary progression of life. Although this is the most modern theory of evolution, it too has major fundamental flaws, which also disqualify it as a realistic evolutionary mechanism.

For example, if mutations are indeed instrumental in the presumed evolutionary progression of life, they should tend to increase the viability and systematization of the organism in which they occur. In reality, however, mutations are almost always (99.99%) harmful, if not lethal, to the unfortunate organism in which they occur. In other words, mutations produce organisms that are weaker and at a marked disadvantage; they are less able to compete for survival. This fact directly contradicts the assumptions and hopeful expectations of the modern evolutionary theory. Incidentally, mutations do not account for the giraffes's long neck either. The slight differences in neck lengths are now known to be caused by differences in food; or by variation in the number of dominant genes that control neck length. [44]

Mutations are not only harmful, but they are also very rare. They occur once in about every ten million duplications of a DNA molecule! [60] Furthermore, mutations are random, not directional. Thus, mutations are unpredictable and do not follow any ordered design or plan, as would certainly be expected if the concept of organic evolution is to have any hope at all. Consequently, mere random mutations cannot account for organized directional evolution; they lack the all-important capacity for intelligent design.

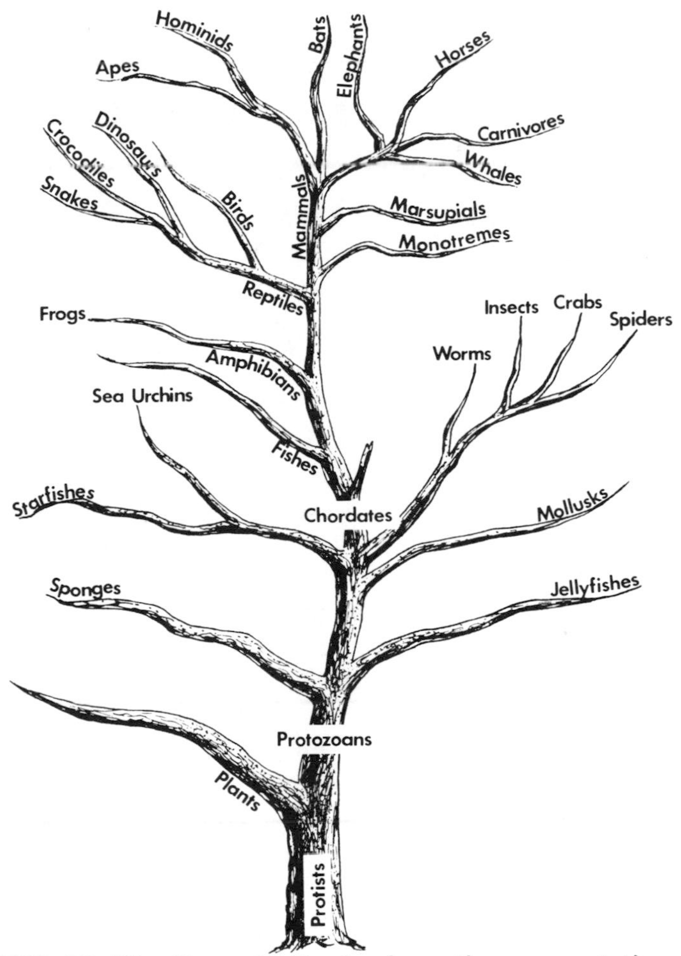

FIG. 18: The Tree of Life. A schematic representation of the supposed evolutionary progression of life. Creationists envision a forest rather than a single tree.

Despite the fact that mutations are known to be 99.99% harmful, very rare, and completely random, Neo-Darwinists nevertheless have faith that these difficulties are sublimated by geologic time and natural selection. They believe that there is enough time to produce and select abundant favorable mutations to be utilized in evolutionary integration.

Although it is true that mere chance processes can produce things, it is equally true that **whatever chance creates, it almost instantaneously annihilates.** Thus, we are not here by a mere chance process for if that were true, we would have vanished long ago by the same objective probability. As it turns out, much to the evolutionist's consternation, time is actually an enemy of organic evolution, not its salvation. The handiwork of time is disassociation and disintegration, not synthesis.

Retreating to the process of natural selection, the Neo-Darwinist still fails to find support for his crumbling theory. Natural selection is only a conservative process that tends to insure the survival and perpetuation of fit organisms, while at the same time eliminating unfit organisms. It is not an innovative process that produces novel structures. It only acts to conserve structures and organisms already in existence.

In conclusion, the Neo-Darwinian combined mechanism of mutation, natural selection, and time cannot account for the presumed evolutionary progression of life. Such a mechanism is, to put it kindly, inadequate for such a monumental task. Thus, today's evolutionist is left without a mechanism for his theory. How, then, can such a person possibly continue to believe in evolution? Is it not for the fact that the god of this world (Satan) has blinded the minds of them who have chosen not to believe the Truth (2 Corinthians 4:4)?

G. DNA

The most intriguing, unique, and important characteristic of all life is the fascinating ability to reproduce.

Reproduction and the transmission of hereditary information (inheritance) are directed by the highly complex DNA (deoxyribo-nucleic acid) molecule. All life, from the most basic bacteria to man, is dependent upon the fantastic DNA molecule. Therefore, if the evolutionary concept is valid, it must have involved the accidental synthesis of a DNA molecule. Frank Salisbury, an evolutionary biologist, discusses the difficulty surrounding the chance development of DNA as follows:

> "Now we know that the cell itself is far more complex than we had imagined. It includes thousands of functioning enzymes, each one of them a complex machine itself. Furthermore, each enzyme comes into being in response to a gene, a strand of DNA. The information content of the gene (its complexity) must be as great as that of the enzyme it controls.
>
> "A medium protein might include about 300 amino acids. The DNA gene controlling this would have about 1,000 nucleotides in its chain, one consisting of 1,000 links could exist in $4^{1,000}$ different forms. Using a little algebra (logarithms) we can see that $4^{1,000} = 10^{600}$. Ten multiplied by itself 600 times gives the figure "1" followed by 600 zeros! This number is completely beyond our comprehension." [66]

A further complication involved with the original synthesis and subsequent replication of the DNA molecule is that the DNA molecule can only be replicated with the assistance of specific enzymes which, in turn, can only be produced by the controlling DNA molecule. Each is absolutely necessary for the other and both must be present for replication to occur. Dr. Haskins, president of the Carnegie Institute of Washington, has commented on this difficulty:

> "Did the code and the means of translating it appear simultaneously in evolution? It seems almost incredible that any such coincidence could have occurred, given the extraordinary complexities of both sides and the requirement that they be coordinated accurately for survival. By a pre-Darwinian (or a skeptic of evolution after Darwin) this puzzle would surely have been interpreted as the most powerful sort of evidence for special creation." [29]

Advances in our understanding of biochemical processes also lend strong support to special creation. Our modern understanding of DNA, proteins, and natural chemical processes, for example, argue powerfully against naturalistic evolution. As Parker has noted:

> "There is no natural chemical tendency for a series of bases to line up a series of R-groups in the orderly way required for life. The base-R group relationship has to be *imposed from the outside.* " [60]

Thus, the hereditary code is not the result of mere chance aided by time and natural processes. It clearly reflects the handiwork of a planned and purposed special creation.

In the human body, DNA "programs" all characteristics such as hair, skin, eyes, and height. DNA determines the arrangement for 206 bones, 600 muscles, 10,000 auditory nerve fibers, 2 million optic nerve fibers, 100 billion nerve cells, 400 billion feet of blood vessels and capillaries, and so on. Further, the capacity of DNA to store information vastly exceeds that of modern technology. The information needed to specify the design of all the species of organisms which have ever lived could be held in a teaspoon and there would still be room left for all the information in every book ever written (Denton, Michael, Evolution: A Theory in Crisis, Adler & Adler, Publishers, Inc., Bethesda, Maryland, 1985, p. 334). Such extraordinary sophistication can only reflect intelligent design.

Furthermore, computer scientists have demonstrated conclusively that information does not and cannot arise spontaneously. [45] Information results only from the expenditure of energy (to arrange the letters and words) and under the all-important direction of intelligence. Therefore, since DNA is information, the only logical and reasonable conclusion that can be drawn is that DNA was formed by intelligence.

The intricately-ordered structure of the DNA molecule is truly an engineering wonder, almost beyond comprehension. This amazing biochemical system could never have arisen apart from Divine creation. Without question, DNA remains one of the greatest testimonies of special creation. As the Psalmist exclaims:

> "I will praise thee; for I am fearfully and wonderfully made: marvelous are thy works; and that my soul knoweth right well."
>
> (Psalm 139:14)

H. Summary

The bewildering complexity, diversity, beauty, order, and astonishing perfection of life forms thoroughly defy evolutionary explanation. Fascinating interrelationships such as mimicry, symbiosis, parasitism, and so on, among organisms clearly indicate purposeful, intelligent design. The widespread existence of amazingly sophisticated migratory and other instincts do not lend themselves to evolutionary elucidation. The numerous instances of convergence and double convergence preclude evolutionary rationale. The incomprehensible complexity of the DNA molecule can only be explained in terms of special creation. And most importantly, the theory of organic evolution lacks a mechansim! Thus, the consistent and overwhelming witness of biology is in support of Biblical creationism and soundly against mythological evolution.

5.

ANTHROPOLOGY

A. Introduction

The origin of man is an issue of extreme importance. The opinions of creationists and evolutionists could hardly be more divergent than on this particular concept. Is man merely the product of an innate evolutionary process, or is he a special creation fashioned directly by his Creator for a specific purpose? Major Biblical principles and concepts such as sin, salvation, accountability, righteousness, eternal judgment, and so on, take on paramount importance or else seem totally irrelevant to an individual, depending on the viewpoint taken.

Evolutionists believe that man and apes evolved from a common unknown ancestor about 30 to 70 million years ago. They point to a number of fossils of bones and teeth to support this claim. Creationists, on the other hand, maintain that man was supernaturally created, completely distinct from all other living creatures (Genesis 1; 1 Corinthians 15:39) only a few thousand years ago. Creationists argue that the fossils cited by evolutionists do not represent stages of human evolution at all, but rather are derived from apes, men, or neither. They are not from animals intermediate between men and apes.

There are many impressive museum exhibits throughout the world that claim to demonstrate human evolution. Although these exhibits are based upon extremely fragmentary evidences, they are nevertheless presented as a well-established fact (Figure 19). For instance, in 1980 the Natural History Museum constructed a permanent exhibition of human evolution. Although there is no direct fossil evidence of man's supposed ancestry, this exhibit strongly conveys the impression that the evolution of man from animals is a fact not to be questioned.

The so-called *ape-men,* which allegedly link men and apes, are named after the place where they are discovered. For example, there is Nebraska Man, Java Man, Peking Man, and so on. We shall now investigate some of the more significant so-called fossil *ape-men* to evaluate the veracity of human evolution.

B. Nebraska Man

Nebraska Man was discovered in 1922 by Harold Cook in the Pliocene deposits of Nebraska. A tremen-

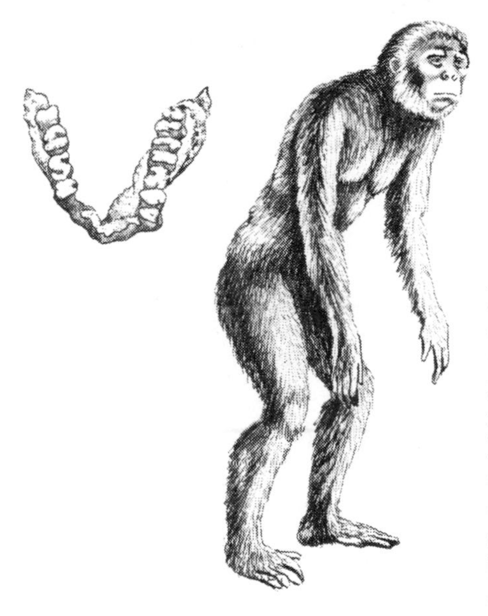

FIG. 19: Ramapithecus. Based upon upper and lower jaws and the teeth of some 30 animals, a complete ape known as Ramapithecus is reconstructed. Such imaginative reconstruction, which is based upon such limited fossil evidence, can only be directed by a preconditioned evolutionary bias, the end product constituting dangerous fiction.

dous amount of literature was built around this supposed missing link which allegedly lived 1 million years ago.

The evidence for Nebraska Man was used by evolutionists in the famous Scopes evolution trial in Dayton, Tennessee in 1925. William Jennings Bryan was confronted with a battery of "great scientific experts" who stunned him with the "facts" of Nebraska Man. Mr. Bryan had no retort except to say that he thought the evidence was too scanty and to plead for more time. Naturally, the "experts" scoffed and made a mockery out of him. After all, who was he to question the world's greatest scientific authorities?

But what exactly was the scientific proof for Nebraska Man? The answer is a tooth. That's right; he found one tooth! The top scientists of the world examined this tooth and appraised it as proof positive of a prehistoric race in America. What a classic case of excessive imagination!

Years after the Scopes trial, the entire skeleton of the animal from which the initial tooth came was found. As it turns out, the tooth upon which Nebraska Man was constructed belonged to an extinct species of pig. **The "authorities," who ridiculed Mr. Bryan for his supposed ignorance, created an entire race of humanity out of the tooth of a pig!** What an embarrassment to the scientific community and a noteworthy commentary on our human nature. Needless to say, little publicity was given to the discovered error. Surely, there is a lesson here for us concerning the reliability of so-called "expert testimony," which is so often used to manipulate and intimidate the layman.

A similar discovery, which was also based upon a tooth, was the Southwest Colorado Man. It is now known that this particular tooth actually belonged to a horse! [17]

How resourceful and imaginative scientific "experts" can be at times. Give them one tooth, not necessarily human, and they can create an entire race of prehistoric humanity.

C. Java Ape-Man

One of the most famous of all the anthropoids is the Java Ape-Man, *Pithecanthropus erectus* (erect ape-man). He was discovered in 1891 by Dr. Eugene Dubois, a fervent evolutionist. Dr. Dubois' find consisted of a small piece of the top of a skull, a fragment of a left thigh-bone, and three molar teeth. Although this evidence is admitedly more substantial, it is still fragmentary. Furthermore, these remnants were not found together. They were collected over a range of about 70 feet. Also, they were not discovered at the same time, but over the span of one year. To further complicate matters, these remains were found in an old river bed mixed in with the bones of extinct animals. Despite all of these difficulties, evolutionists calmly assure us that Java Ape-Man lived about 750,000 years ago.

Although the "experts" would have us to believe that these mere fragments provide sufficient information from which to reconstruct an entire prehistoric race, certain questions are raised. For instance, how is it possible to reconstruct so completely with such confidence from such scanty evidence? How can the "experts" be so certain that all the pieces came from the same animal? How have these unpetrified bones managed to survive for so long without disintegrating? And so on.

Well, as it turns out, even the "experts" differed greatly about the identification of these fossil fragments. In fact, of the twenty-four European scientists who met to evaluate the find, ten said they came from an ape; seven from a man; and seven said they belonged to a no longer missing link. Controversy and division surrounded the discovery. The renowned Professor Virchow of Berlin said:

> "There is no evidence at all that these bones were parts of the same creature." [17]

Even Dr. Dubois himself later reversed his own opinion. His final conclusion was that the bones were the remains of some sort of gibbon. But one would never gather the truly equivocal nature of the world-famous Java Ape-Man by viewing museum exhibits or reading college textbooks, which are so dogmatic. The dubious nature of Java Ape-Man (and human evolution as well) is either conveniently ignored or concealed behind the mask of *"scientific fact."*

One final note regarding Java Ape-Man. Another *Pithecanthropus* was found in Java in 1926. Typically, this discovery was also billed as a prodigious breakthrough, the missing link for sure. It turned out to be the knee-bone of an extinct elephant. [17]

D. Piltdown Man

The remains of Piltdown Man were allegedly discovered in 1912 by Charles Dawson, an amateur fossilologist. He produced some bones, teeth, and primitive implements, which he said he found in a gravel pit at Piltdown, Sussex, England. He took them to Dr. Authur Smith Woodward, an eminent paleontologist at the British Museum. The remains were acclaimed by anthropologists to be about 500,000 years old. A flood of literature followed in response to this discovery with Piltdown Man being hailed in the museums and textbooks as the most wonderful of finds. Over 500 doctoral dissertations were performed on Piltdown Man. [61] Surely, this find will stand the test of time and establish evolution as a fact of science; or will it?

All was well until October of 1956 when the entire hoax was exposed. Reader's Digest came out with an article, summarized from *Popular Science Monthly*, entitled The Great Piltdown Hoax. Using a new method to date bones based upon fluoride absorption, the Piltdown bones were found to be fraudulent. Further critical investigation revealed that the jaw-bone actually belonged to an ape that had died only 50 years previously. The teeth were filed down, and both teeth and bones were discolored with bichromate of potash to conceal their true identity.[17] And so, Piltdown Man was built upon a deception which completely fooled all the "experts" who promoted him with the utmost confidence. According to M. Bowden:

> "... the person responsible for placing the faked fossils in the pit at Piltdown was Teilhard de Chardin S.J."[10]

Teilhard authored several philosophical books in which he attempted to harmonize evolution and Christianity. Exasperated by the lack of convincing evidence for Darwin's theory, Teilhard was apparently motivated into *assisting* the theory of evolution by fabricating the needed missing link.

It should be noted that Piltdown Man was viewed in stately museums and studied in major textbooks for several generations. What will today's "facts" of human evolution turn out to be in the near future? And so, once again, the veracity of "expert testimony" is called into question. How fitting are the words of Scripture which declare:

> "Professing themselves to be wise, they became fools."
>
> (Romans 1:22)

E. Neanderthal Man

Neanderthal Man was first discovered at about the turn of the century in a cave in the Neanderthal Valley near Dusseldorf, Germany. He was portrayed as a semi-erect, barrel-chested, brutish sort of fellow, an intermediary link between man and apes.

With the discovery of other neanderthal skeletons, it is now known, however, that Neanderthal Man was fully erect and fully human. In fact, his cranial capacity even exceeded that of modern man by more than 13%.[78]

The old misconceptions about Neanderthal Man were due to two factors: first, the bias of the pre-programmed evolutionary anthropologists who reconstructed him; and second, the fact that the particular individual on whom the initial evaluation was made was crippled with osteo-arthritis and rickets. Today Neanderthal Man is classified as *Homo sapiens,* completely human.[25]

F. Lucy

Present-day speculation about human evolution revolves about a group of fossils called australopithecines and, in particular, a specimen called *Lucy,* a 40% complete skeleton. Lucy was discovered by D.C. Johanson in the Afar area of Ethiopia during investigations conducted from 1972-1977.

In a *National Geographic* article (December 1976) Johanson claimed that:

> "The angle of the thigh bone and the flattened surface at its knee joint end . . . proved she walked on two legs."[33]

However, the knee joint end of the femur was severely crushed; therefore, Johanson's conclusion is pure speculation. Anatomist Charles Oxnard, using a computer technique for analysis of skeletal relationships, has concluded that the australopithecines did not walk upright, at least not in the same manner as humans. In this connection, it should be mentioned that the chimpanzee spends a considerable amount of time walking upright. Thus, there is no valid scientific basis for a conclusion of bipedalism in Lucy. Lucy and her relatives are probably just varieties of apes.

Finally, there is evidence that people walked upright before the time of Lucy. This would include the Kanapoi hominid and Castenedolo Man. Obviously, if people walked upright before the time of Lucy, then she must be disqualified as an evolutionary ancestor.[61]

G. Summary

In closing, we note the highly speculative, unreliable, and imaginative nature of anthropology. In each case, mere fragments are the basis for reconstructing these so-called "ape-men." Preconditioned artists, guided by their imaginative evolutionary bias, reconstruct the needed "missing links," determining posture, expression, stature, and so on. Sweeping claims are made, and a flood of favorable literature typically follows such discoveries, all based upon a few fossil fragments.

To base the proposed evolutionary ancestry of man upon such fragmentary evidence is highly questionable and very misleading especially when the evidence is not honestly presented. The colossal blunders and outright hoaxes of the past certainly support this fact. The Bible tells us that some "...shall turn away their ears from the truth, and shall be turned unto fables" (2 Timothy 4:4); and again that some are "ever learning, and never able to come to the knowledge of the truth" (2 Timothy 3:7).

Creationists contend that the only true record of man's ancestry, which began with Adam and Eve, is that which is recorded in the Bible. The Bible speaks clearly of man as a special creation, entirely unrelated to the animal kingdom by any sort of evolutionary connection (Genesis 1; 1 Corinthians 15:39). Far from being an evolutionary accident of nature, man is the crown of creation, made in the very image of God (Genesis 1:26-27).

". . .keep that which is committed to thy trust, avoiding profane and vain babblings, and oppositions of science falsely so called." (1 Timothy 6:20).

6.

COMMONLY CITED "PROOFS" OF EVOLUTION

A. Introduction

In order to support their theory, evolutionists often point to certain so-called "proofs" of evolution, examples which they believe actually demonstrate organic evolution. Typically, these examples are presented in a dogmatic fashion as proof-positive of evolutionary integration. In this chapter we shall examine several of the most commonly cited "proofs" including the famous fossil horse series, vestigial organs, the peppered moth, the duck-billed platypus, Archaeopteryx, the Biogenetic "Law," the Miller-Urey experiment, and comparative anatomy. We shall endeavor to show that these "proofs" are actually mere *assumptions,* and that the evidence is more hypothetical than empirical.

B. Fossil Horse Series

One of the most highly praised and well-known examples of "proof" for organic evolution is the famous fossil horse series (Figure 20). Far from being the well-established fact, which evolutionists portray it to be, the horse series is plagued with many serious problems and is actually nothing more than a deceitful delusion. Consider the following list of major difficulties and discrepancies with the fossil horse series:

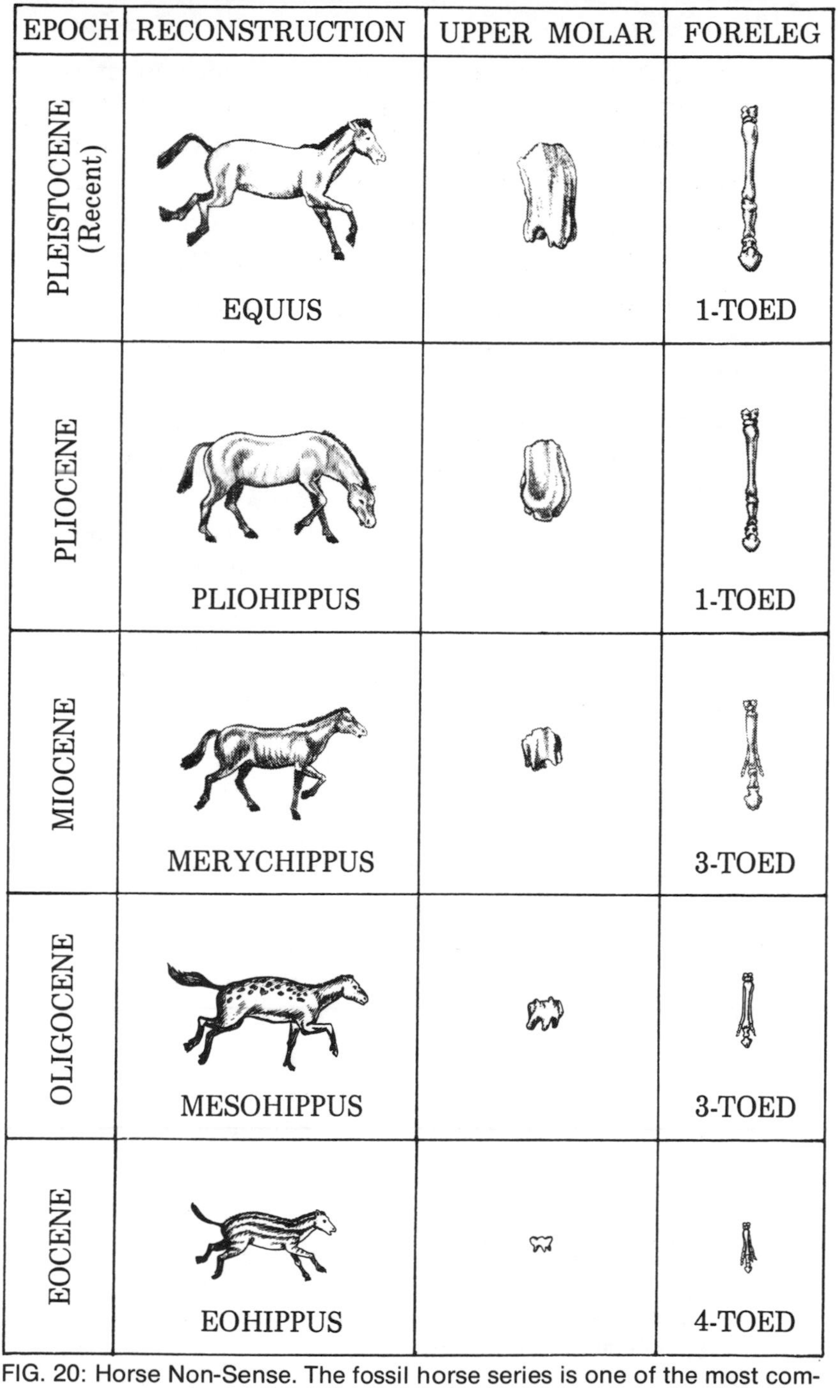

EPOCH	RECONSTRUCTION	UPPER MOLAR	FORELEG
PLEISTOCENE (Recent)	EQUUS		1-TOED
PLIOCENE	PLIOHIPPUS		1-TOED
MIOCENE	MERYCHIPPUS		3-TOED
OLIGOCENE	MESOHIPPUS		3-TOED
EOCENE	EOHIPPUS		4-TOED

FIG. 20: Horse Non-Sense. The fossil horse series is one of the most commonly cited evidences in support of evolution. It is, however, plagued with numerous major difficulties and discrepancies. It has been constructed on the basis of evolutionary presupposition rather than on scientific fact.

1. A complete series of horse fossils in the correct evolutionary order does not exist anywhere in the world. [36]

2. There are over 20 different genealogical trees of the so-called fossil horse series. [11]

3. The fossil horse series starts in North America, jumps to Europe, and then back again to America.[36]

4. The sequence from small many-toed forms to large one-toed forms is completely absent in the fossil record.[78]

5. Eohippus, the earliest member of the horse evolution series, is completely unconnected by any sort of link to its presummed ancestors, the condylarths. [36]

6. There are no evolutionary intermediates between each of the horses. Each appears abruptly in the fossil record. [78]

7. The teeth of the animals found are either grazing or browsing types. There are no transitional types of teeth. [44]

8. Two modern-day horses (*Equus nevadenis* and *Equus occidentalis*) have been found in the same fossil stratum as Eohippus! [78] This fact is fatal to the concept of horse evolution since horses were already horses before their supposed evolution.

The so-called fossil horse series may actually represent different horse genera. They may have been members or variations of originally created Biblical kinds. In any event, there is certainly no valid scientific reason for assuming that the horses evolved rather than were created as distinct kinds, which lived at the same time. [36] As George Gaylord Simpson has written:

> "The uniform continuous transformation of Hyracotherium into Equus, so dear to the hearts of generations of textbook writers, never happened in nature." [68]

C. Vestigial Organs

Vestigial organs are those structures which are presumed by evolutionists to be the useless remains of an organ which was once fully developed and operational in ancestral types. Such structures have long been cited as evidence for evolution since they are assumed to represent former evolutionary changes.

As it turns out, however, the case of vestigial organs may actually be an example of the evolutionist speaking too soon. Advances in our understanding of physiology have shown that supposed vestigial organs are actually quite useful and even essential. For instance, textbooks as recent as the 1960's listed over 200 vestigial structures for the human body, including the thyroid and pituitary glands! As our knowledge and understanding of these organs and structures has increased, the list of "useless" structures and organs has decreased. Today all organs formerly classed as vestigial are known to have some function during the life of the organism.

The fatal flaw in the argument from vestigial organs is exposed by modern genetics. Basically, the concept of vestigial organs represents a return to Lamarckism where the development or loss of a structure is based upon need. It is now known, however, that organs can only be altered by a genetic alteration in the chromosomes, or DNA. The use or disuse of an organ has no effect whatsoever on subsequent generations.

Even if the concept of vestigial organs were valid, it still would not lend support to evolution since it implies structures on the way out, not in. Nascent organs, those under construction into a functional unit, are completely nonexistent. This fact serves as a powerful argument against organic evolution.

In conclusion, the entire concept of vestigial organs is biologically indefensible and to be completely discarded as the misconception of a former day.

D. Peppered Moths

Evolutionists commonly cite the case of the peppered moth (*Biston betularia*) of England as a striking example of present-day Neo-Darwinian evolution. Peppered moths have always existed in light, intermediate, and dark-colored varieties. Before the advance of the industrial revolution, the tree trunks were light-colored and the light-colored moths were camouflaged; whereas, the dark-colored moths were easily spotted and eaten by birds. Consequently, the dark-colored moths constituted a very minor proportion of the total population.

As the industrial revolution progressed, however, and pollution increased, the tree trunks became darker and within 45 years the situation was reversed. In the Manchester vicinity, for example, 95% of the moths were of the dark-colored variety.

But is this really evolution? Certainly not! This process did not produce anything new. It did not result in increased complexity and organization. The dark-colored moths had always existed. The air pollution simply caused a shift in populations of the dark versus light-colored moths. Although absolutely no evolutionary change occurred in these moths, the case of the peppered moths does illustrate the principle of natural selection. In the introduction of the 1971 edition of the Origin of Species, L. Harrision Matthews notes:

> "The (peppered moths) experiments beautifully demonstrate natural selection—or survival of the fittest—in action, but they do not show evolution in progress, for however the populations may alter in their content of light, intermediate, or dark forms, all the moths remain from the beginning to end *Biston betularia.*" [42]

Despite these obvious facts, many textbooks and encyclopedias continue to cite the peppered moth as an example of evolutionary development. The International

Wildlife Encyclopedia, for instance, refers to this case as:

> ". . . the most striking evolutionary change ever to be witnessed by man." [13]

But if this is the best example that evolutionists can offer to substantiate their theory, then they are indeed in serious trouble for this is not evolution at all.

E. Duck-Billed Platypus

Evolutionists insist that the duck-billed platypus is an evolutionary link between mammals and birds. The platypus is an Australian mammal which has fur and nourishes its young with milk, like mammals. Their young hatch from eggs like reptiles, and they also have webbed feet and a flat bill like a duck. It has pockets in its jaws to carry food and a spur on its rear legs which, like a snake's fang, is poisonous. And, amazingly enough, the platypus even uses echo location like dolphins! Admitedly, the platypus does combine a curious mixture

FIG. 21: Duck-Billed Platypus. The Duck-billed Platypus possesses a curious mixture of traits. The evolutionist's allegation that it is a transitional form is not supported by the facts. The Platypus appears to be a distinct kind of animal that has been specifically designed to include a mosaic of traits.

of features and traits (Figure 21). But does this unusual combination of characteristics necessarily mean that the platypus is a transitional creature? Creationists reply, "absolutely not" and argue that the platypus is simply a kind of organism whose design includes a mosaic of traits.

Although it might seem as though the duck-billed platypus could be used to argue either point of view, there are several good reasons for rejecting the evolutionary interpretation of the origin of the platypus. A few of these reasons include:

1. Platypus fossils are exactly the same as modern forms.

2. The complex structures of the egg and milk glands are always fully developed and offer no solution as to the origin and development of the womb or the milk glands.

3. The more typical mammals are found in much lower strata than the egg-laying platypus. [8]

Thus, the duck-billed platypus appears to be a distinct kind of animal in and of itself that has been specifically designed to include a mixture of traits. Rather than being an evolutionary transitional form, the duck-billed platypus is yet another demonstration of our Creator's virtuosity.

F. Archaeopteryx

Evolutionists tell us that birds have evolved from reptiles. However, the only fossil ever found and proposed to be a transitional link between these two classes is the famous Archaeopteryx (Figure 22). It is difficult to understand (if you believe in evolution) why only one fossil has ever been uncovered that might be transitional between these two groups of different animals especially since evolutionists estimate an eighty million year developmental time span from reptiles to birds.

(a) The fossil imprint from the Jurassic Solnhofen limestone in Germany; (b) An artist's reconstruction of Archaeopteryx.

FIG. 22: Archaeopteryx. Once hailed by evolutionists as a striking example of a reptile-bird, Archaeopteryx is today classified by most scientists as a true bird. This is because each of the characteristics of Archaeopteryx is found either in true birds, or not found in reptiles.

Where are the millions and millions of other intermediary forms, which demonstrate the evolution of flight from the presumed reptilian ancestor? Where are the prized fossils of specimens consisting of half scales and half feathers? After all, if the evolution of reptiles into birds is indeed a reality, vast numbers of transitional forms should be objectively preserved in the fossil record, and yet they are not.

Furthermore, although it has been argued that Archaeopteryx combined certain reptilian and avian characteristics, it is today classified by most paleontologists as a true bird, not a reptile-bird intermediary. This is because each of the features of Archaeopteryx is either found to exist in true birds or be absent in many reptiles. [78] Archaeopteryx appears to be no more a link between reptiles and birds than the duck-billed platypus is a link between mammals and birds.

Another serious difficulty with the supposed evolution of reptiles into birds is that which concerns their lungs. The lungs of reptiles consist of millions of tiny air sacs; whereas, bird's lungs have tubes. The piecemeal evolution of bird's lungs from reptile's lungs seems virtually impossible. The survival of the hypothetical intermediate life forms possessing lungs, which consist of half tubes and half air sacs, is totally inconceivable. [36]

Finally, fossils of modern birds have been found in the same rocks as Archaeopteryx. This means that Archaeopteryx cannot have been the forerunner of birds since birds were already in existence. Thus, concerning evolutionary transitional links, Archaeopteryx is completely irrelevant.

G. The Biogenetic "Law"

In 1866, Ernst Haeckel advanced the so-called *Biogenetic Law.* Also known as the *Recapitulation Theory,* this concept is still presented in college textbooks. It can be summarized in the popular cliché, "Ontogeny recapitulates phylogeny." Simply stated, this means that the development of an embryo (ontogeny) is supposed to retrace the imagined evolutionary development (phylogeny) of the organism.

A favorite example used by evolutionists to support the Biogenetic "Law" is the human heart. Allegedly, the human heart passes through a worm, fish, frog, and reptile stage before achieving the final human stage. Structures resembling fish gills in the human embryo were also said to demonstrate a fish ancestry for man.

Modern research has, however, exposed the numerous fallacies of this "law." For instance, researchers have demonstrated conclusively that the various stages through which an embryo passes are essential for its development from a single cell into a highly complex and ordered organism. Furthermore, countless exceptions, reversals, omissions, and additions in the embryologic sequences have been observed in embryologic studies. Finally, the field of molecular genetics establishes the impossibility of the Biogenetic "Law." DNA is very specific and uniquely programmed for each type of organism. It simply does not recreate passing developmental stages of other organisms. It only produces after its own kind.

Almost all scientists now reject the Biogenetic "Law." Only naive or poorly informed evolutionists still cite this concept in defense of their theory for it has no valid scientific foundation whatsoever.

H. The Miller-Urey Experiment

The origin of life is an area of great interest and intense research today. Research centers have been established worldwide to investigate the origin of life, and many experiments have been conducted in an attempt to duplicate supposed spontaneous origins.

The most well-known and highly acclaimed experimental model is that of Stanley Miller and H.C. Urey. In their experiment, water vapor, ammonia, methane, and hydrogen were subjected to spark discharges, and simple amino acids were synthesized. Although this experiment demonstrates conclusively that organic compounds can be artificially made, such products do not even remotely approach the synthesis of life. Also, achieving this sort of accomplishment in the laboratory under carefully designed and controlled conditions, and having the exact same thing (or something more complex) occur in an open system apart from any directing intelligence, are two entirely different things.

There are a number of serious objections to the Miller-Urey chemical evolution experiment, which render the relevancy of the model as suspect at best. A few of these objections are as follows:

1. The concentrations of methane and ammonia were carefully selected to insure the production of organic molecules. There is no evidence to suggest that the earth's primitive atmosphere was so characterized.[78]

2. There is no evidence to indicate that the earth's early atmosphere was reducing. There is, however, considerable evidence to suggest that the earth had an oxidizing atmosphere over most, if not all, of its history.[74]

3. A methane-ammonia reducing atmosphere would be fatal to life forms.[78]

4. The simulation of lightning by mild spark discharges is unrealistic. Actual lightning would have destroyed any organics which may have been present.[78]

5. The molecules produced in the Miller-Urey apparatus would react detrimentally to life forms which were trying to evolve. Chemically, they would destroy all hope of producing life.[60]

To the creationist, the fact that it takes the world's greatest intellects utilizing the most advanced technology to unravel the most basic secrets of life only serves to indicate the necessity for intelligence in the beginning.

Man's childish attempts to create life are completely futile. In fact, even if scientists were able to produce all of the necessary chemicals in the required relationships, life still would not result. Anyone who would like to verify that statement has only to try and raise the dead. After all, the dead already contain all the complex ingredients scientists are attempting to generate (cells, DNA, enzymes, and so on). Wysong makes this point quite clear and extinguishes the hopes and dreams of all chemical evolution experimentors:

> "Is not the hope to produce life by arriving at proper chemicals and chemical relationships in experiments a rather bizarre aspiration? It is futile. The trillions of organisms that die and have remained dead stand as solemn proof that life is a unique supernatural quality not reproducible by the puny efforts of men wistfully reaching like moths for the stars." 78

I. Comparative Anatomy

The science of comparative anatomy deals with the study of the physical structures of animals. Structural similarities among various unrelated animals are considered by evolutionists to be a strong evidence for organic evolution. They argue that homologous structures (those which are similar in structure, but not necessarily in function) prove a common ancestry of animals (Figure 23).

Creationists do not deny the existence of similarity in design or structure, which comparative anatomy certainly reveals. Creationists do, however, take issue with the evolutionist's interpretation of this phenomenon. The Biblical interpretation of homologous structures argues for special creation according to a common basic blueprint designed by one Master Architect. The general pattern was, of course, flexible enough to accomodate variations and modifications on the basic plan to enable various kinds of organisms to fit in their pre-determined enviroments. In other words, there was no need or reason to make a completely different pattern or blueprint for every species when the basic pattern was perfectly satisfactory for a variety of functions.

Although it might be well argued that both interpretations are only based upon subjective preconceived beliefs, the creationist's viewpoint is more empirical. Consider, for example, the fact that both the fossil record and the living world reveal a great variety of organisms, which are distinctly set apart from one another. If the evolutionary interpretation were correct we should find a continuous and gradual inter-gradation between all life forms back to their supposed common ancestor. The fact that no such fine gradation exists refutes the evolutionary interpretation and supports the creationist's viewpoint, which predicts both similarities and differences as commonly observed in nature.

Another difficulty with the use of so-called *homolo-*

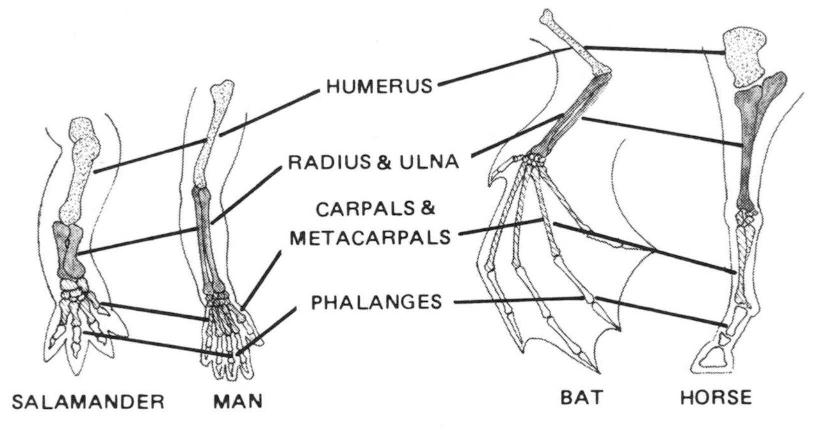

FIG. 23: Comparative Anatomy. Evolutionists argue that homologous structures suggest a common ancestry but this allegation is not supported by the facts and does not appear to be the correct interpretation. Creationists view such similarities in terms of a Creator who chose to use a common pattern in His creative designs.

gous structures to support the evolutionary philosophy is that the same genes that produced the presumed ancestral structure also produce the present corresponding structure. In other words, the genes involved in producing the structure are the same, but the structure itself has changed. Why is this a problem for evolutionists? Simply because many so-called *homologous structures* are produced under the direction of entirely different genes.

J. Summary

To summarize, the commonly cited *proofs* of evolution—the fossil horse series, vestigial organs, the peppered moth, the duck-billed platypus, Archaeopteryx, the Biogenetic "Law," the Miller-Urey experiment, and comparative anatomy—are either mere unfounded assumptions or known erroneous concepts. Since evolution has not been evidenced in the fossil record and is not observed today, evolutionists are completely deprived of any empirical evidence with which to support their dubious hypothesis. The tenuous nature of evolution is by now most apparent, its collapse already a reality in the minds of the well-informed.

7. SCIENTISTS WHO BELIEVE IN BIBLICAL CREATIONISM

It is commonly asserted and accepted that all "real" scientists believe in organic evolution. Those who do not, regardless of degrees, position, and so on, are simply not true scientists. This arrogant and plainly prejudiced view is completely false and easily refuted.

Consider the fact that many of the greatest pioneering scientists of the past were Bible-believing creationists. A partial listing would certainly include such august men as: Louis Agassiz, Francis Bacon, Robert Boyle, George Cuvier, Henri Fabre, Michael Faraday, John Ambrose Flemming, William Herschel, James Joule, Lord Kelvin, Johann Kepler, Carolus Linnaeus, Joseph Lister, James Maxwell, Gregor Mendel, Samuel F.B. Morse, Issac Newton, Blaise Pascal, Louis Pasteur, Lord Raleigh, and Leonardo daVinci![47] Creationists are certainly in good company (See Appendix C).

Today there are literally thousands of highly reputable scientists representing every scientific discipline who are Biblical creationists. Their numbers continue to grow and many creation-science research centers and organizations have been formed (See Appendix B).

Many of today's most distinguished scientists completely dismiss the concept of organic evolution in favor of Biblical creationism. For instance, Dr. John Grebe, director of basic and nuclear research for Dow Chemical Company, is offering $1,000 to anyone who can produce just one clear proof of evolution. Dr. Grebe's challenge is not to be taken lightly. His credentials are extremely impressive, having over 100 patents and being responsible for the development of styrofoam, synthetic rubber, and Saran Wrap. [7]

Co-holder of the 1945 Nobel Prize for developing penicillin, Sir Ernest Chain, has recently stated:

> "To postulate that the development and survival of the fittest is entirely a consequence of chance mutations seems to me a hypothesis based on no evidence and irreconcilable with the facts. These classical evolutionary theories are a gross over-simplification of an immensely complex and intricate mass of facts, and it amazes me that they are swallowed so uncritically and readily, and for such a long time, by so many scientists without a murmur of protest." [65]

The 1971 winner of the Nobel Peace Prize in science, Dr. Dennis Gabor, has stated:

> "I just cannot believe that everything developed by random mutations . . . " [7]

Dr. Etheridge, world-famous paleontologist of the British Museum, has remarked:

> "Nine-tenths of the talk of evolutionists is sheer nonsense, not founded on observation and wholly unsupported by facts. This museum is full of proofs of the utter falsity of their views. In all this great museum, there is not a particle of evidence of the transmutation of species." [41]

Dr. Albert Fleischmann, of the University of Erlangen,

has written:

> "I reject evolution because I deem it obsolete; because the knowledge, hard won since 1830, of anatomy, histology, cytology, and embryology, cannot be made to accord with its basic idea. The foundationless, fantastic edifice of the evolution doctrine would long ago have met with its long-deserved fate were it not that the love of fairy tales is so deep-rooted in the hearts of man."[1]

Many other highly distinguished scientists of our day could be quoted who completely reject the evolutionary philosophy. These should suffice, however, to refute the evolutionist's allegation that Bible-believing scientists are second-rate, pseudo-scientists. Numerous important breakthroughs, discoveries, and inventions were accomplished by creationist scientists and many of the greatest scientific minds, which the world has ever known, were and are Biblical creationists, who completely reject and scorn the scientifically-bankrupt evolutionary philosophy.

8.

BIBLICAL CONSIDERATIONS

The preceeding chapters have shown that the facts of science are better explained in terms of Biblical creationism and catastrophism rather than by atheistic evolutionism and uniformitarianism. The limitless diversity, complexity, and order of the universe clearly indicate and testify of a wonderful Master Designer and Creator. True science supports and confirms the Genesis account of creation and repudiates organic evolution. In this final chapter, we will direct our attention toward Biblical considerations as related to the atheistic evolutionary philosophy.

Many Bible-believing Christians have been intimidated and beguiled by the widespread evolutionary philosophy into compromising their Christian beliefs and embracing unscriptural ideologies. But for those who properly revere the Bible as the unadulterated Word of God, there is no room nor need for compromise. Recognizing that there are numerous major blatant contradictions between the Biblical account of creation and evolution, the devoted Bible-believing Christian can only reject the evolutionary philosophy in its entirety. **The two concepts are totally incompatible; they are mutually exclusive.** Consider the following list of contradictions which clearly demonstrate the impossibility of reconciling evolution with Scripture:

1. **Bible:** God is the Creator of all things (Genesis 1).
   Evolution: Natural chance processes can account for the existence of all things.

2. **Bible:** World created in six literal days (Genesis 1).
   Evolution: World evolved over the aeons.

3. **Bible:** Creation is completed (Genesis 2:3).
   Evolution: Creative processes continuing.

4. **Bible:** Oceans before land (Genesis 1:2).
   Evolution: Land before oceans.

5. **Bible:** Atmosphere between two hydrospheres (Genesis 1:7).
   Evolution: Contiguous atmosphere and hydrosphere.

6. **Bible:** First life on land (Genesis 1:11).
   Evolution: Life began in the oceans.

7. **Bible:** First life was land plants (Genesis 1:11).
   Evolution: Marine organisms evolved first.

8. **Bible:** Earth before sun and stars (Genesis 1:14-19).
   Evolution: Sun and stars before earth.

9. **Bible:** Fruit trees before fishes (Genesis 1:11, 20, 21).
   Evolution: Fishes before fruit trees.

10. **Bible:** All stars made on the fourth day (Genesis 1:16).

Evolution: Stars evolved at various times.

11. **Bible:** Birds and fishes created on the fifth day (Genesis 1:20, 21).
    Evolution: Fishes evolved hundreds of millions of years before birds appeared.

12. **Bible:** Birds before insects (Genesis 1:20-31; Leviticus 11).
    Evolution: Insects before birds.

13. **Bible:** Whales before reptiles (Genesis 1:20-31).
    Evolution: Reptiles before whales.

14. **Bible:** Birds before reptiles (Genesis 1:20-31).
    Evolution: Reptiles before birds.

15. **Bible:** Man before rain (Genesis 2:5).
    Evolution: Rain before man.

16. **Bible:** Man before woman (Genesis 2:21-22).
    Evolution: Woman before man (by genetics).

17. **Bible:** Light before the sun (Genesis 1:3-19).
    Evolution: Sun before any light.

18. **Bible:** Plants before the sun (Genesis 1:11-19).
    Evolution: Sun before any plants.

19. **Bible:** Abundance and variety of marine life all at once (Genesis 1:20-21).
    Evolution: Marine life gradually developed from a primitive organic blob.

20. **Bible:** Man's body from the dust of the earth (Genesis 2:7).
    Evolution: Man evolved from monkeys.

21. **Bible:** Man exercised dominion over all organisms (Genesis 1:28).
    Evolution: Most organisms extinct before man evolved.

22. **Bible:** Man originally a vegetarian (Genesis 1:29).
    Evolution: Man originally a meat-eater.

23. **Bible:** Fixed and distinct *kinds* (Genesis 1:11, 12, 21, 24, 25; 1 Corinthians 15:38-39).
    Evolution: Life forms in a continual state of flux.

24. **Bible:** Man's sin the cause of death (Romans 5:12).
    Evolution: Struggle and death existent long before the evolution of man. [48,54]

In addition to these specific direct contradictions, there are stark differences of general principle between atheistic evolution and Biblical Christianity. Jesus said:

> "A good tree cannot bring forth evil fruit, neither can a corrupt tree bring forth good fruit."
>
> (Matthew 7:18)

The fruit of evolution has been all sorts of anti-Christian systems of beliefs and practice. It has served as an intellectual basis for Hitler's nazism and Marx's communism. It has promoted apostasy, atheism, secular humanism, and libertinism as well as establishing a basis for ethical relativism, which has spread through our society like a cancer. The mind and general welfare of mankind has suffered greatly as a result of this naturalistic philosophy.

According to the Bible, man is a responsible creature. One day he will give an account for his life's actions and motives. But when man is viewed as the product of some vague purposeless evolutionary process, he is conveniently freed from all moral obligations and responsibilities. After all, he is merely an accident of nature, an intelligent animal at best. His bad behavior is simply an excusable remnant of his animal history. If one continues this line of reasoning to its logical conclusion, the panacea for mankind would be found in his ability to change and control his environment. Consequently, those who believe in evolution are forced to look to social planners for their salvation, rather than to their only real hope—Jesus Christ. Man's nature cannot be truly changed from the outside in, but can only come from an inward change, which will correct the outward. As Jesus said: " . . . Ye must be born again" (John 3:1-21).

Another area in which evolution has been especially influential in perverting the minds of men is in morality. Biblical morality, exemplified by the teachings and very life of Jesus Christ, is without question, the ultimate standard of morality. But if chance is our creator, then no universal code of moral absolutes exists. Our present-day *new morality,* which would be more accurately labeled *no morality,* is the inevitable consequence of such an atheistic philosophy of origins. It is no mere coincidence that the modern deterioration of morality has occurred contemporaneously with the advance of the evolutionary philosophy. The moral collapse of today can indeed be largely attributed to the very system that wipes out all moral standards. Of course, this is one reason for the popularity of the evolutionary philosophy. The only solution to this deplorable trend is a genuine repentance in the hearts of men and a sincere return to Biblical standards and principles.

Many other Biblical difficulties with the theory of evolution could be listed. The foregoing considerations, however, should suffice to establish the fact that Bible-believing Christians must reject any and all forms of evolutionary thinking. Compromising theistic evolution simply cannot and need not be reconciled with Biblical Christianity.

> " . . . ye shall know the truth, and the truth shall make you free."
>
> (John 8:32)

CONCLUSION

It is impossible to make observations or conduct experiments when dealing with the subject of origins. Therefore, both evolution and Biblical creationism cannot be proven scientifically to be true. They belong in the realm of beliefs. Each is a religious philosophy of origins, which requires faith (perhaps *credulity* would be a more accurate term in the case of evolution). This is especially true of evolution as Dr. Wysong has observed:

> "Evolution requires plenty of faith: a faith in L-proteins that defy chance formation; a faith in the formation of DNA codes which if generated spontaneously would spell only pandemonium; a faith in a primitive environment that in reality would fiendishly devour any chemical precursors to life; a faith in experiments that prove nothing but the need for intelligence in the beginning; a faith in a primitive ocean that would not thicken but would only hopelessly dilute chemicals; a faith in natural laws including the laws of thermodynamics and biogenesis that actually deny the possibility for the spontaneous generation of life; a faith in future scientific revelations that when realized always seem to present more dilemmas to the evolutionist; faith in probabilities that

treasonously tell two stories—one denying evolution, the other confirming the creator; faith in transformations that remain fixed; faith in mutations and natural selection that add to a double negative for evolution; faith in fossils that embarrassingly show fixity through time, regular absence of transitional forms and striking testimony to a world-wide water deluge; a faith in time which proves to only promote degradation in the absence of mind; and faith in reductionism that ends up reducing the materialist's arguments to zero and forcing the need to invoke a supernatural creator." [78]

Far from being the established fact of science that it is so typically portrayed to be, evolution is, in reality, an unreasonable and unfounded hypothesis that is riddled with countless scientific fallacies. Biblical creationism, on the other hand, does correlate with the known facts of science. Indeed, there is an abundance of impressive and convincing scientific evidence which can be used to justify an intelligent faith in Biblical Christianity.

The widespread influence of evolution is largely responsible for our moral decline of recent years. Belief in atheistic evolution has led many individuals to reject the Biblical account of creation, and along with that, the entire message of the Bible itself. Herein lies the awesome danger of this Satanic delusion—many will fail to receive the grace of God in the Gospel of Jesus Christ, which leads to salvation and eternal life. Instead, the followers of naturalistic evolution can only experience a meaningless existence on earth and a hopeless future. As the Bible cautions:

> "There is a way which seemeth right unto a man; but the end thereof are the ways of death."
>
> (Proverbs 16:25)

But the message of the Bible is good news, the Gospel

of Jesus Christ! Herein is the offering for the forgiveness of sins, purpose and meaning for life, and the sure promise of everlasting life in Heaven with God to all who respond in faith (John 3:16).

The collapse of evolution, then, is already a reality in the minds and hearts of the well-informed. The ever-accumulating weight of scientific and Biblical fact has finally crushed the superficial fraud of organic evolution. Refusal to accept the validity of Biblical Christianity can no longer be based upon a lack of factual evidence. The amount of evidence in support of Biblical creationism and catastrophism is indeed copious. Such rejection must be viewed as being of the *will* and not the *mind*; not so much "I cannot believe," but "I will not believe." Alleged intellectual difficulties with Biblical Christianity are usually nothing more than a smoke screen for moral rebellion against God and His Word.

The erroneous concept of organic evolution is ephemeral, here today and gone tomorrow, destined to pass into obscurity. Therefore, place your faith this very day in your Creator and Savior, the One who supersedes time and all else—changing not:

> "Jesus Christ the same yesterday, and today, and for ever."
>
> (Hebrews 13:8)

APPENDIX A

Scientific Facts Which Prove Evolution

Listed on this page are all of the known scientific facts which can be used to prove that evolution is an established fact of science, as commonly taught:

APPENDIX B

Creation-Science Organizations

AUSTRALIA

Creation-Science Association
6PO Box 2935
Adelaide, South Australia 5001

Creation-Science Foundation
Cnr. Bradman & Bellrick
Acacia Ridge, Queenland
Australia 4110

Creation-Science Foundation Limited
P.O. Box 302
Sunnybank Queensland
Australia 4109

Creation-Science Movement
"Bethuel" 13 Beddows St.
Burwood, Victoria
Australia 3125

BRAZIL

Associacao Brasileira
de Pesquisa da Criaca
Caixa Postal 37, 36570
Vicosa, MinaGeraise
Brazil

CANADA

Creation Science Association of Alberta
P.O. Box 9075 Stat. E
Edmonton, Alberta
Canada T5P 4K1

Creation Science Association of Canada
P.O. Box 367
Surrey, British Columbia
Canada V3T 5B6

Creation Science Association of Manitoba
P.O. Box 68
Rosenort, Manitoba
Canada

Creation Science Association of Ontario
298 Hillcrest Ave.
Willowdale, Ontario
Canada M2N 3P4

Creation Science of Saskatchewan
Box 26
Kenaston, Saskatchewan
Canada S0G 2N0

International Christian Crusade
205 Yonge St. Room 31
Toronto
Canada V8N 1C5

The North American Creation Movement
1556 Arrow Road
Victoria, British Columbia
Canada V8N 1C5

ENGLAND
Biblical Creation Society
51 Cloan Crescent
Bishopbriggs, Glasglow
United Kingdom G64 2HN

Creation News
3 Church Terrace
Cordiff, United Kingdom
CF2 5AW

Creation Science Movement
Rivendell 20 Foxley Lane
High Salvington, Worthing
England BN13 3AB

Creation Science Movement
50 Brecon Ave.
Cosham
Portsmouth PO6 2AW

The Creation-Science Movement
40 Chesford Groud.
Stratford-Upon-Avon
Warwickshire CV379LS

Reseach Scientists
Christian Fellowship
38 DeMontfort St.
Leicester, England LEI 7GP

Research Scientists
Christian Fellowship
39 Bedford Square
London, England
WCI B 3EY

Somerset Creationist Group
Mead Farm, Downhead, West Camel
Yeovil, BA 22 7RQ
Somerset, England

GERMANY

Studiengemeinschaft Wort and Wissen e.v.
Zumberger See 91
D-5820 Gevelsberg
Germane

INDIA

Creation Research
Rajpayga Road
Gwalior, M.P., 474009
India

KOREA
Korea Association of Creation Research
15-5 Jung Dong, Jung-Ku
Seoul 100, Korea

MEXICO
Ciencia y Creation
Apartado No. 1759-A
Chihuahua, Chihuanua,
Mexico

NETHERLANDS
Stichting tot Bevordering van Bybelgetrowe
Wetenschap (Foundation for the Advancement of
Studies Faithful to the Bible)
Secretary: Kampweg 106
3941 HL Doorn, Netherlands

NEW ZEALAND
Creation Literature Society, Inc.
48 Craig Road, Maraetai Beach, Auckland
New Zealand

SOUTH AFRICA
Creation Science Movement
Ms. S. Wood
210 Kennedy Street
Kenilworth, Johannesburg,
South Africa 2190

SPAIN
Coordinadora Creacionista
Apartado 2002
Sabadell, (Barcelona) Spain

SWEDEN
Association for Christian Belief
Okome, Prastgard Pl. 4703
31060 Ullared Sweeden

Forening for Biblisk
Skapelestro
Box 3170, 400 10 Goteborg
Sweden

UNITED STATES

American Scientific Affiliation
5 Douglas Avenue
Elgin, IL 60120

Association for Biblical Research
P.O. Box 31
Willow Grove, PA 19090

Baltimore Creation Fellowship, Inc.
P.O. Box 356
Perry Hall, MD 21236

Bible Science Association
P.O. Box 33187
Granada Hills, CA 91344

Bible Science Association
2911 E. 42nd Street
Minneapolis, MN 55409

Bible-Science Association
1255 Poplar Avenue
P.O. Box 3624
Memphis, TN 38173

Bible-Science Association
101 North Superior
Oscoda, MI 48750

Bible-Science Association
10926 Hole
Riverside, CA 92505

Bible-Science Association
P.O. Box 66507
Seattle, WA 98166

Bible-Science Association
7663 Wentworth Avenue
Tujunga, CA 91042-1636

Bible-Science Association
1429 Holyoke
Wichita, KS 67208

The Bible-Science Center
360 Fairfield Road
Fairfield, NJ 07006

Bible-Science Studies
6304 N. College Avenue
Oklahoma City, OK 73122

Center for Creation Studies
Liberty Baptist College
Lynchburg, VA 24506

Center for Scientific Creation
1319 Brush Hill Circle
Naperville, IL 60540

Central Ohio Creation-Science Association
1007 Groveport Road
Canal Winchester, OH 43110

Citizens for Fairness in Education
1516 Danube Lane
Plano, TX 75075

Citizens for Scientific Creation
P.O. Box 164
Saratoga, CA 95070

Creation Education Association
Eugene A. Sattler
Rt. 1, Box 161
Pine River, WI 54965

Creation Filmstrip Centre
Route 1, Box 94
Haviland, KS 67059

Creation Health Foundation
19 Gallery Centre
Taylors, SC 29687

The Creation Report
805 Meadow Lane
Plover, WI 54467

Creation Research Science Education Foundation
P.O. Box 292
Wheaton, IL 60189

Creation Research Society
P.O. Box 14016
Terre Haute, IN 47803

Creation Science Association
18346 Beverly Rd.
Birmingham, MI 48009

Creation Science Association
2825 Riva Ridge Circle
Cottage Grove, WI 53527

Creation Science Association of Michigan
26938 Northmore
Dearborn, MI 48127

Creation Science Association of Middle Tennessee
P.O. Box 972
Brentwood, TN 37027

Creation Science Association of Orange County
P.O. Box 4325
Irvine, CA 92716-4325

Creation Science Committee
P.O. Box 2282
Fairbanks, AK 99707

Creation Science Fellowship
362 Ashland Avenue
Pittsburg, PA 15228

Creation Science for Mid-America
Route 1, Box 247B
Cleveland, MO 64734

Creation Science Foundation
P.O. Box 6064
Evanston, IL 60202

Creation Science Legal Defense Fund
1200 N. Market Street, Suite J
Shreveport, LA 71107

Creation Science Research Center
P.O. Box 23195
San Diego, CA 92123

Creation Science Society
201 S. Brent St.
Ventura, CA 93003

Creation Science Society of Milwaukee
5334 N. 66th St.
Milwaukee, WI 53218

El Paso Bible-Science Association
37 Sacramento Ave.
El Paso, TX 79930

Fair Education Foundation
Rte. 2, P.O. Box 415
Murphy, NC 28906

Fellowship of Saved Scientists & Interested Laymen
77 Baker St.
Jamestown, NY 14701

GEOSCIENCE RESEARCH INSTITUTE
Loma Linda University
Loma Linda, CA 92350

The Institute for Creation Research
10946 Woodside Ave.
Santee, CA 92071

K.V. Creationists Club
430 Kentucky
Lawrence, KS 66044

Lutheran Research Society
2222 B. Street
Forest Gove, OR 97116

The Lutheran Science Institute
357 East Howard Avenue
Milwaukee, WI 53207

Midwest Creation Fellowship
Box 75
Wheaton, IL 60189

Missouri Association for Creation
405 North Sappington Road
Glendale, MO 63122

Museum of Creation Pub.
1457 Washington
Lincoln, NE 68502

National Foundation for Fairness in Education
P.O. Box 1
Washington, D.C. 20044

Northcoast Bible-Science Association
7594 Biscayne
Parma, OH 44134

The Origins Research Society
7756 East Mariposa Drive
Scottsdale, AZ 85251

Rochester Creation Science Association
61 Thistledown Drive
Rochester, NY 14617

The Scientific Educational Foundation
734 Greenwood
Toledo, OH 43605

Silicon Valley Bible-Science Association
13445 Harper Drive
Saratoga, CA 95070

South Bay Creation Science Association
22322 Harbor Ridge Lane, #2
Torrance, CA 90502

Students for Origin Research
P.O. Box 203
Goleta, CA 93116

Twin Cities Creation Science Association
1136 5th Avenue, South
Anoka, MN 55303

APPENDIX C

Scientific Disciplines and Contributions by Creationist Scientists

Discipline/Contribution	Scientist
Absolute Temperature Scale	Lord William Kelvin (1824-1907)
Antiseptic Surgery	Joseph Lister (1827-1912)
Bacteriology	Louis Pasteur (1822-1895)
Barometer	Blaise Pascal (1623-1662)
Biogenesis Law	Louis Pasteur (1822-1895)
Calculating Machine	Charles Babbage (1792-1871)
Calculus	Isaac Newton (1642-1727)
Celestial Mechanics	Johannes Kepler (1571-1630)
Chemistry	Robert Boyle (1627-1691)
Chloroform	Sir James Simpson (1811-1870)
Classification System	Carolus Linnaeus (1707-1778)
Computer Science	Charles Babbage (1792-1871)
Dimensional Analysis	Lord John Rayleigh (1842-1919)
Double Stars	Sir William Herschel (1738-1822)
Dynamics	Sir Isaac Newton (1642-1727)
Electric Generator	Michael Faraday (1791-1867)
Electric Motor	Joseph Henry (1797-1878)
Electronics	Sir John A. Fleming (1849-1945)
Electro-Magnetics	Michael Faraday (1791-1867)
Entomology of Living Insects	Jean Henri Fabre (1823-1915)
Fluid Mechanics	Sir George Stokes (1819-1903)
Galactic Astronomy	Sir William Herschel (1738-1822)
Galvanometer	Joseph Henry (1797-1878)
Gas Dynamics	Robert Boyle (1627-1691)
Genetics	Gregor Mendel (1822-1884)
Glacial Geology	Louis Agassiz (1807-1873)
Gynecology	Sir James Simpson (1811-1870)
Hydraulics	Leonardo da Vinci (1452-1519)
Inert Gases	Sir William Ramsay (1852-1916)
Law of Gravity	Sir Isaac Newton (1642-1727)
Non-Euclidean Geometry	Bernhard Riemann (1826-1866)
Oceanography	Matthew Maury (1806-1873)
Optical Mineralogy	Sir David Brewster (1781-1868)
Paleontology	John Woodward (1665-1728)
Pasteurization	Louis Pasteur (1822-1895)
Pathology	Rudolf Virchow (1821-1902)
Physical Astronomy	Johannes Kepler (1571-1630)
Reflecting Telescope	Isaac Newton (1642-1727)

Discipline/Contribution	Scientist
Reversible Thermodynamics	James Joule (1818-1889)
Scientific Method	Lord Francis Bacon (1561-1626)
Statistical Thermodynamics	James C. Maxwell (1831-1879)
Stratigraphy	Nicolaus Steno (1638-1686)
Systematic Biology	Carolus Linnaeus (1707-1778)
Telegraph	Samuel F. B. Morse (1791-1872)
Thermodynamics	Lord William Kelvin (1824-1907)
Thermokinetics	Sir Humphry Davy (1778-1829)
Transatlantic Cable	Lord William Kelvin (1824-1907)
Vaccination & Immunization	Louis Pasteur (1822-1895)
Vertebrate Paleontology	Georges Cuvier (1769-1832)

APPENDIX D

Glossary

abyssal - Of or pertaining to the great depths of the sea, generally below 1,000 fathoms (6,000 feet).

adaptation - An imagined evolutionary process through which an organism acquires characteristics which make it better suited to live and reproduce in its environment.

allele - One of a pair of characters that are alternative to each other in inheritance, being governed by genes situated at the same locus in homologous chromosomes.

amphibian - A vertebrate of the class Amphibia, presumed by evolutionists to be intermediate between the fishes and reptiles, including frogs, toads, newts, salamanders, and so on.

Ancestral - Connected with; derived from; having character of, ancestors.

antediluvian - Before the time of the Genesis Flood.

anthropoids - The group of animals that evolutionists believe are closely related to man; chimpanzee, gorilla, gibbon, and so on.

anthropology - The science of man, especially with relation to his physical character, the origin and distribution of races, human environment and social relations, and culture.

aperture - An opening; hole, gap.

Archaeopteryx - A famous fossil of an extinct bird once believed by evolutionists to be a transitional form between reptiles and birds.

assumption - An unprovable belief used as a basis for reasoning.

astronomic time - The assumed age of the universe; approximately 30 billion years according to evolutionary scientists.

Aves - The class of vertebrates comprising the birds.

avian - A bird-like characteristic.

Biogenetic "Law" - Also known as the Recapitulation Theory, this "law" maintained that the development of an embryo retraces the evolutionary development of the organism. Coined by the phrase "Ontogeny recapitulates phylogeny," this "law" is no longer accepted by scientists.

Biology - The science of life.

bipedalism - The ability to walk on two feet.

catastrophe - A sudden, violent change in the physical conditions of the earth's surface; a cataclysm.

catastrophism - The belief that the fossils, rock formations, and other features of the earth's crust have been formed rapidly in a relatively short period of time during a worldwide disaster; i.e., the Genesis Flood.

chemical evolution - The imagined formation of complex chemical compounds from simpler compounds and elements by purely natural processes before any life existed on the earth.

chromosome - Any of the microscopic rod-shaped bodies into which the chromatin seperates during mitosis: they carry the genes that convey the hereditary characteristics and are constant in number for each species.

coal - An impure form of carbon found in the forms of lignite, bituminous, and anthracite.

Coelacanth - A supposedly long-extinct crossopterygian fish once thought to be a transitional form between fishes and amphibians. It is now known that this fish is still alive today as one was caught near Madagascar in 1938.

cold-blooded - Animals having blood that varies in temperature with the surrounding environment; fishes, reptiles, and so on.

comet - A heavenly body having a starlike nucleus with a luminous mass around it, and usually a long, luminous tail; comets move in orbits around the sun.

comparative anatomy - The science that deals with the study of the physical structures of organisms.

conglomerate - A detrital sedimentary rock made up of more or less rounded fragments of such size that an appreciable percentage of the volume of the rock consists of particles of granule size or larger.

convergence - Also known as parallelism, it is the presumed parallel and independent evolutionary development of similar features in unrelated animals.

cornea - The transparent front of the outer coat of the eyeball.

creationism - The belief that the universe and the things in it were spoken into existence by miraculous acts of God as described in the Book of Genesis, chapters one and two.

creationist - One who believes in the Biblical account of Creation.

cross-bedding - The arrangement of laminations of strata transverse or oblique to the main planes of stratification of the strata concerned.

crystalline lens - A lens-shaped body within the eye which focuses the rays of light.

crystallization - The process through which crystals seperate from a fluid, viscous, or dispersed state.

Darwinism - Also referred to as the Theory of Natural Selection, this idea can be stated simply as "survival of the fittest."

data - Information based on observation and experimentation.

daughter isotope - The isotope created by the radioactive decay of a parent isotope. The amount of a daughter isotope continually increases with time.

delta - A deposit of sediment, roughly triangular, at the mouth of a river.

DNA - Deoxyribose nucleic acid. A nucleoprotein which assures that living organisms will reproduce after their kind.

dolomite - A mineral composed of the carbonate of calcium and magnesium, $CaMg(CO_3)_2$.

dominant - In genetics, designating or of that one of any pair of opposite Mendelian characters which, when factors for both are present in the germ plasm, dominates over the other and appears in the organism; opposite of recessive.

duck-billed platypus - An Australian, oviparous mammal, with a broad flat beak like a duck's, and webbed feet.

ecology - A study of the relationship between organisms and their environment.

energy - the ability or capacity to do work.

entropy - A measure of the quantity of energy not capable of conversion into work.

environment - All the biotic and abiotic factors that actually affect an individual organism at any point in its life cycle.

enzyme - A protein that acts as a catalyst to speed up chemical reactions in living cells.

ephemeral marking - A special type of fossil originally formed as a transient marking on the surface of a recently deposited layer of sediment. These include ripple marks, rain imprints, worm trails, and bird and reptile tracks.

erosion - The process that loosens and moves sediment to another place on the earth's surface. Agents of erosion include water, ice, wind,and gravity.

evaporites - Rocks that result from evaporation of mineralized water. Common examples are rock salt and gypsum.

evolution - An imaginary process by which nature is said to continually improve itself through gradual development.

evolutionist - One who attempts to explain the origin of things by theories of naturalistic, gradual development.

extinct - Not found alive today; found only as a fossil.

extinction - Act of extinguishing; state of being extinct; annihilation.

fact - Something observed or measured; an actual event, occurence, quality, or relation.

factorial - The product of a given series of consecutive integers from 1 on up.

fauna - The aggregation of animal species characteristic of a certain locality, region, or environment. The animals found fossilized in certain geologic formations or occuring in specified time intervals of the past may be referred to as fossil faunas.

first law of thermodynamics - Also known as the law of energy conservation, this law states that energy can be converted from one form into another, but it can neither be created nor destroyed.

fossil - The remains or traces of animals or plants, which have been preserved by natural causes in the earth's crust; evidence of past life.

fossil graveyards - Large groups of fossils that look as if they all died together or were washed into the same burial site.

gene - In genetics, any of the elements in the chromosome by which hereditary characters are transmitted and determined.

genetics - The science of heredity through the genes.

geologic column - An imagined chronological arrangement of rock units in columnar form with the presumed oldest units at the bottom and presumed youngest at the top.

geologic time - The assumed age of the earth; estimated to be about five billion years old by evolutionary scientists.

geologic time-scale - An imagined chronological sequence of units of earth time.

geology - The branch of science dealing with the physical nature, history, and development of the earth.

gibbon - A small, slender, long-armed ape of India, southern China, and the East Indies.

greenhouse effect - Short waves of light entering an enclosure, absorbed by objects, and re-emitted at higher wavelengths so that they cannot escape from the enclosure.

half-life - The amount of time required for half of a radioactive isotope to decay to another isotope of less mass.

historical geology - The study of the origin and development of the earth and its inhabitants.

homologous structures - Similar anatomical structures in many plants and animals that appear to be unrelated.

horizontal variation - The normal variation permitted within the range specified by the DNA for a particular type of organism. No novel characteristics, which exhibit higher degrees of order or complexity, are produced. An example of horizontal variation is the great variety of different types of dogs.

host - A mineral that contains an inclusion.

host rock - The wall rock of an epigenetic ore deposit.

hydraulic - Pertaining to fluids in motion.

in situ - In its original place.

integration - An imagined evolutionary process through which less ordered systems give rise to more ordered systems, resulting in increasingly higher and higher levels of complexity.

invertebrates - Animals that do not have a hard backbone, such as worms and starfish.

iris - The opaque circular curtain, or diaphragm, forming the colored part of the eye, perforated by the pupil.

Lamarckism - Also known as the Theory of Inheritance of Acquired Characteristics, this theory proposes that changes occuring in an organism could be passed on to its offspring. This theory is rejected by scientists today.

lignite - A brownish-black coal in which the alteration of vegetal material has proceeded further than in peat but not so far as sub-bituminous coal.

Linnaeus, Carolus (1707-1778) - Swedish botanist and naturalist who established the first orderly system of classification for all living things (kingdom, phylum, class, order, family, genus, species). The universal usage of binomial nomenclature to indicate a genus and species is also credited to Linnaeus.

lithification - The processs by which unconsolidated rock-forming materials are converted into a consolidated or coherent state.

mammals - Animals with hair or fur whose young are nursed on milk.

marine - Refers to the salt water environment.

mathematics - The science concerned with the properties of, and relations between quantities.

micrometeoroids - Small inter-planetary dust particles.

mimicry - The superficial resemblance that some animals exhibit to other animals or natural objects, thereby securing concealment or protection.

mutation - An inheritable change in the chromosomes; usually a change of a gene from one allelic form to another.

nascent organs - Imaginary organs that are under construction supposedly evolving into something useful.

natural selection - The process by which those individuals with characteristics that help them become adapted to their environment tend to survive (survival of the fittest) and transmit their characteristics.

Neo-Catastrophism - The idea that major flooding (not the Genesis Flood) offers the most reasonable explanation for the fossil record.

Neo-Darwinism - The modern synthetic theory of evolution; Darwinism modified in accordance with the principles of genetics.

niche - The role or profession of an organism in the environment; its activities and relationships in the community.

nuclear fusion - The fusion of atomic nuclei, as of heavy hydrogen or tritium, into a nucleus of heavier mass, as of helium, with a resultant loss in the combined mass, which is converted into energy: the principle of the hydrogen bomb.

nucleus - The central core of some atom models containing all of the positive charge.

optic - Pertaining to the eye or to vision.

oxidizing atmosphere - An atmosphere containing a substantial amount of oxygen so that any chemical reactions would normally be of substance with oxygen.

ozone - The allotropic form of oxygen having three atoms per molecule. Ozone is represented as O_3.

ozone layer - The ozone-containing portion of the stratosphere, which absorbs harmful ultraviolet radiation from the sun.

paleontology - The study of ancient plant and animal life based on fossil remains found in the earth; the study of fossils.

petrified - A condition when minerals dissolved in ground-water replace the cells of a dead tree or bone, and produce a mineral copy of the organism.

petroleum - An oily, liquid mixture of many hydrocarbons.

Phlogiston Theory - The erroneous belief that combustible materials contain a substance called phlogiston, said to be released as the substance burns.

physics - The science dealing with the properties, changes, interaction, and so on of matter and energy.

physiology - The branch of biology dealing with the functions and vital processes of living organisms or their parts and organs.

pleochroic halo - A minute, concentric spherical zone of darkening or coloring that forms around inclusions of radioactive minerals in biotite, chlorite, and a few other minerals; about 0.075mm in diameter.

polystrate fossils - Fossils, such as tree trunks, that extend through many sedimentary layers.

Poynting-Robertson Effect - An astronomic process of the sun through which small inter-planetary dust particles are eliminated either by being pushed into space by radiation pressure or by being swept into the sun.

primordial - The state of existing at or from the beginning; original.

probability - The likelihood of an event.

punctuated equilibrium - The supposition that evolution occurs by sudden and large leaps rather than through gradual, small changes; originated by evolutionists to help account for the lack of intermediate forms in the fossil record.

pupil - The small opening in the center of the iris which appears like a black spot in the middle of the eye. It contracts or expands according to the degree of light.

quadruped - An animal, especially a mammal, with four feet.

radioactive dating - The process of finding an apparent age of a sample based on the amount of radioactive substance present in the material.

reagent - A substance used to detect or measure another substance or to convert one substance into another.

recessive - In genetics, designating or of that one of any pair of opposite Mendelian characters which, when factors for both are present, remains latent; opposite of dominant.

reducing atmosphere - An atmosphere containing substances that would prevent oxidation reactions. Oxides would be reduced or decomposed in a reducing atmosphere.

retina - The inner membrane of the eyeball containing the light-sensitive rods and cones, which receive the optical image.

salvation - The saving of one's soul from the eternal consequences of sin, as by repentance and faith in the atonement of Jesus Christ.

second law of thermodynamics - Also known as the law of energy decay, this law states that every system left to its own devices always tends to move from order to disorder, its energy tending to be transformed into lower levels of availability.

sedimentary rocks - Those rocks that have been formed by the accumulation of sediment in water (aqueous deposits) or from air (aeolian deposits).

Seymouria - A fossil animal once believed to have been a transitional form between amphibians and reptiles.

Silica - Silicon dioxide, SiO_2.

spontaneous generation - The false belief that life arose from nonliving matter. This idea was disproved by Louis Pasteur.

strata - Layers of sedimentary rock.

symbiosis - The living together of two or more organisms of different species in a mutually beneficial relationship.

synthesis - Putting together; the formation of a compound by the combining of two or more simpler compounds, elements, or radicals.

teleology - Purposiveness or design in nature as an explanation of natural phenomena.

theistic evolution - The attempted harmonization of evolution with a belief in God.

tidal zone - Generally considered to be the zone between mean high-water and mean low-water levels.

topography - The shape and form of the earth's surface.

transitional forms - Forms of life that are supposed to show how one type of trait changed into another as one kind evolved into another.

transmutation - The imagined evolutionary transformation of one form into a completely different form.

trilobite - An extinct animal with three body lobes, jointed legs, a crab-like outer skeleton, and sometimes eyes.

uniformitarianism - The concept that the present is the key to the past. Processes now operating to modify the earth's surface are believed to have been operating similarly in the geologic past; that there is a uniformity of processes past and present.

vapor - The gaseous state of substance.

vertebrates - Animals with hard backbones, such as birds and mammals.

vertical variation - The imagined variation potential within organisms which enables them to develop into higher, more complex, and entirely different kinds of organisms.

vestigial organs - Structures that were once presumed by evolutionists to have been the useless remains of an organ, which was once fully developed and operational in ancestral types.

vitreous humor - A transparent, jelly-like substance, which fills the hinder portion of the eyeball.

water vapor - Water in the gaseous state.

weathering - The chemical and mechanical breakdown of rock.

REFERENCES

[1]Acworth, Captain Bernard, "Darwin and Natural Selection," *Evolution Protest Pamphlet,* London, 1960, p. 6.

[2]Ayala, Francisco J., "Teleological Explanations in Evolutionary Biology," *Philosophy of Science,* Vol. 37, March 1970, p. 3.

[3]Baker, Sylvia, *Bone of Contention: Is Evolution True?,* Evangelical Press, England, 1980, p. 33.

[4]Barnes, Thomas G., "Depletion of the Earth's Magnetic Field," *Impact No. 100,* Institute for Creation Research, California, October 1981, p. 4.

[5]Barnes, Thomas G., *Origin and Destiny of the Earth's Magnetic Field,* Institute for Creation Research, Technical Monograph No. 4, Creation-Life Publishers, Inc., San Diego, California, 1973, p. 12.

[6]Barnes, Thomas G., "Young Age for the Moon and Earth," *Impact No. 110,* Institute for Creation Research, California, August 1982, p. 4.

[7]Blick, Edward F., *Special Creation Vs. Evolution,* Southwest Radio Church, PO Box 1144, Oklahoma City, Oklahoma, 1981, p. 38.

[8]Bliss, Richard B., Gary E. Parker, and Duane T. Gish, *Fossils: Key to the Present,* Creation-Life Publishers, Inc., San Diego, California, 1980, p. 81.

[9]Blum, Harold F., *Time's Arrow and Evolution,* Princeton University Press, Princeton, N.J., 1962, p. 119.

[10] Bowden, M., *Ape-Men – Fact or Fallacy?,* Sovereign Publications, Bromley, Kent, Canada, 1977, pp. 35, 46-47.

[11]Bowden, M., *The Rise of the Evolution Fraud,* Creation-Life Publishers, San Diego, California, 1982, p. 117.

[12]Bowden, M., and J. V. Collyer, "Quotable Quotes for Creationists," *Creation Science Movement,* Pamphlet No. 228, January 1982, p. 1.

[13]Burton, M., and R. Burton, Eds., *The International Wildlife Encyclopedia,* Marchal Cavendish Corp., New York, 1970, p. 2706.

[14]Butt, Stephen M., "Insect Flight: Testimony to Creation," *Creation Science Movement,* Pamphlet No. 226, July 1981, p. 4.

[15]Clark, R.E.D., *Christian Belief and Science,* Muhlenberg Press, Philadelphia, Pennsylvania, 1960, pp. 64-74.

[16]Colbert, E., *Evolution of the Vertebrates,* Wiley, New York, 1955.

[17]Criswell, W.A., *Did Man Just Happen?,* Zondervan Publishing Co., Grand Rapids, Michigan, 1973, p. 87.

[18]Curtis, Helena, *Biology,* Worth Publishers, Inc., New York, New York, 1969, p. 862.

[19]Darwin, Francis, Ed., *The Life and Letters of Charles Darwin,* Vol. 1, p. 210.

[20]Darwin, Charles, *The Origin of Species,* Vol. 2, 6th Ed., p. 49.

[21]Davies, L. Merson, "Scientific Discoveries and Their Bearing on the Biblical Account of the Noachian Deluge," *Journal of the Transactions of the Victoria Institute,* LXII, 1930, pp. 62-63.

[22]Dillow, Joseph C., *The Waters Above: Earth's Pre-Flood Vapor Canopy,* Moody Press, Chicago, 1981, p. 479.

[23]Eddington, A.S., *The Nature of the Physical World,* Macmillan, New York, 1930, p. 74.

[24]Gish, Duane T., *Dinosaurs: Those Terrible Lizards,* Creation-Life Publishers, Inc., San Diego, California, 1976, p. 62.

[25]Gish, Duane T., *Evolution: The Fossils Say No!,* Creation-Life Publishers, Inc., San Diego, California, 1976, p. 122.

[26]Gish, Duane T., and Richard B. Bliss, "Summary of Scientific Evidence for Creation," *Impact No. 95,* Institute for Creation Research, California, May 1981, p. 4.

[27]Golay, Marcel, "Reflections of a Communications Engineer," *Analytical Chemistry,* Vol. 33, June 1961, p. 23.

[28]Gould, S.T., and N. Eldredge, "Punctuated Equilibria: The Tempo and Mode of Evolution Reconsidered," *Paleobiology,* Vol. 3, 1977, p. 147.

[29]Haskins, Caryl P., "Advances and Challenges in Science in 1970," *American Scientist,* Vol. 59, May-June 1971, p. 298.

[30]Hitching, Francis, "Was Darwin Wrong?", *Life Magazine,* Vol. 5, No. 4, April 1982, pp. 48-52.

[31]Huber, B, "Recording Gaseous Exchange Under Field Conditions," *The Physiology of Forest Trees,* Ronald Publishers, New York, 1958.

[32]Ingersole, Zobel, and Ingersoll, *Heat Conduction with Engineering, Geological, and Other Applications,* University of Wisconsin Press, 1954, pp. 99-107.

[33]Johanson, Donald, "Ethiopia Yields First "Family" of Early Man," *National Geographic Magazine,* December 1976, Vol. 150, No. 6, pp. 790-811.

[34]Keith and Anderson, "Radiocarbon Dating: Fictitious Results with Mollusk Shells," *Science,* Vol. 141, 1963, p. 634.

[35]Kelvin, Lord, *Journal of the Victoria Institute,* Vol. 124, p. 267.

[36]Kofahl, R.E., *Handy Dandy Evolution Refuter,* Beta Books, San Diego, California, 1977, p. 27.

[37]Lammerts, W.E., *Why Not Creation?,* Baker Book House, Grand Rapids, Michigan, 1973, p. 388.

[38]Laughlin "Excess Radiogenic Argon in Pegmatite Minerals," *Journal of Geophysical Research,* Vol. 74, 1969, p. 6684.

[39]Levitt, Z., *Creation: A Scientist's Choice,* Victor Books, Wheaton, Illinois, 1971, p. 131.

[40]Lindsay, Dennis, "The Dinosaur Dilemma," *Christ for the Nations,* Vol. 35, No. 8, November, 1982, pp. 4-5, 14.

[41]Lindsay, Gordon, *Evolution - The Incredible Hoax,* Christ for the Nations, Dallas, Texas, 1977, p. 16.

[42]Matthews, L.H., *The Origin of Species,* (Introduction) by Charles Darwin, J.M. Dent and Sons, Ltd., London, 1971, p. 10.

[43]Melnick, Jim, *The Case of the Polonium Radiohalos,* Students for Origins Research, Santa Barbara, California, Vol. 5, No.1, 1982, pp. 4-5.

[44]Moore, John, N., and Harold S. Slusher, Eds., *Biology: A Search for Order in Complexity,* Zondervan Publishing House, Grand Rapids, Michigan, 1970, p. 403.

[45]Moorhead, P., and M. Kaplan, Eds., *Mathematical Challenges to the Neo-Darwinian Interpretation of Evolution,* Wistar Institute, Philadelphia, Pennsylvania, 1967.

[46]Morris, H.M., and Duane T. Gish, Eds., *The Battle for Creation,* Creation-Life Publishers, San Diego, California, Vol. 2, 1976, p. 321.

[47]Morris, H.M., "Bible-Believing Scientists of the Past," *Impact No. 103,* Institute for Creation Research, California, January 1982, p. 4.

[48]Morris, H.M., *Biblical Cosmology and Modern Science,* Baker Book House, Grand Rapids, Michigan, 1970, p. 146.

[49]Morris, H.M., Duane T. Gish, and George M. Hillestad, Eds., *Creation: Acts, Facts, Impacts,* Creation-Life Publishers, San Diego, California, 1974, p. 188.

[50]Morris, H.M., *The Genesis Record,* Baker Book House, Grand Rapids, Michigan, 1980, p. 677.

[51]Morris, H.M., *Many Infallible Proofs,* Creation-Life Publishers, San Diego, California, 1975, p. 381.

[52]Morris, H.M., "Probability and Order Versus Evolution," *Impact No. 73,* Institute for Creation Research, California, July 1979, p. 1.

[53]Morris, H.M., *The Remarkable Birth of Planet Earth,* Creation-Life Publishers, San Diego, California, 1978, p. 115.

[54]Morris, H.M., *Scientific Creationism,* Creation-Life Publishers, San Diego, California, 1976, p. 277.

[55]Morris, H.M., *The Troubled Waters of Evolution,* Creation-Life Publishers, San Diego, California, 1980, p. 217.

[56]Morris, John, *Tracking Those Incredible Dinosaurs and the People Who Knew Them,* Creation-Life Publishers, San Diego, California, 1980, p. 238.

[57]Nelson, Byron, *After Its Kind,* Bethany Fellowship, Inc., Minneapolis, Minnesota, 1970, p. 202.

[58]Neville, George T., "Fossils in Evolutionary Perspective," *Science Progress,* Vol. 48, January 1960, pp. 1, 3.

[59]Newell, N.D., "Adequacy of the Fossil Record," *Journal of Paleontology,* Vol. 33, May 1959, p. 496.

[60]Parker, Gary E., *Creation: The Facts of Life,* Creation-Life Publishers, San Diego, California, 1980, p. 163.

[61]Parker, Gary E., "Origin of Mankind," *Impact No. 101,* Institute for Creation Research, California, November 1981, p. 4.

[62]Redding, William S., Ed., *The Lincoln Library,* The Frontier Press Company, Columbus, Ohio, 1970, Vol. 2, p. 2207.

[63]Reno, Cora A., *Evolution on Trial,* Moody Press, Chicago, Illinois, 1970, p. 192.

[64]Riegle, D.D., *Creation or Evolution,* Zondervan Publishing House, Grand Rapids, Michigan, 1971, p. 94.

[65]Rosevear, D.T., "Scientists Critical of Evolution," *Evolution Protest Movement,* Pamphlet No. 224, July 1980, p. 4.

[66]Salisbury, Frank B., "Doubts About the Modern Synthetic Theory of Evolution," *American Biology Teacher,* September 1971, pp. 336-338.

[67]Shute, E., *Flaws in the Theory of Evolution,* Craig Press, Nutley, New Jersey, 1961, p. 468.

[68]Simpson, George Gaylord, *Life of the Past,* Yale University Press, New Haven, Connecticut, 1953, p. 119.

[69]Strokes, William, and William Lee, *Essentials of Earth History,* Prentice-Hall Inc., Englewood Cliffs, New Jersey, 1966, p. 468.

[70]Tax, Sol, Ed., "Evolution After Darwin," *Issues in Evolution,* Chicago University Press, 1960, Vol. 3, p. 41.

[71]Von Fange, E., "Time Upside Down," *Creation Research Quarterly,* Vol. 11, 1974.

[72]Whitcomb, John C., Jr., and Henry M. Morris, *The Genesis Flood,* The Presbyterian and Reformed Publishing Co., 1961, p. 518.

[73]Whitcomb, John, C., Jr., *The World that Perished,* Baker Book House, Grand Rapids, Michigan, 1973, p. 155.

[74]White, A.J. Monty, *What About Origins?,* Dunestone Printers Ltd., Great Britain, 1978, p. 170.

[75]Wilder-Smith, A.E., *The Natural Sciences Know Nothing of Evolution,* Master Books, San Diego, California, 1981, p. 166.

[76]Willey, J., *Convergence in Evolution,* London, England, John Marray, 1911.

[77]William, Emmott L., and George Mulfinger, Jr., *Physical Science for Christian Schools,* Bob Jones University Press, Greenville, South Carolina, 1974, p. 628.

[78]Wysong, R.L., *The Creation-Evolution Controversy,* Inquiry Press, Midland, Michigan, 1981, p. 300-301.

[79]Young, Davis A., *Creation and the Flood: An Alternative to Flood Geology and Theistic Evolution,* Baker Book House, Grand Rapids, Michigan, 1977, p. 217.

[80]Young, Davis A., "Genesis: Neither More Nor Less," *Eternity Magazine,* Vol. 33, No. 5, May 1982, pp. 14-19.

SCRIPTURE INDEX

NAME INDEX

SUBJECT INDEX

D

E

Electrons, 28, 68

N

O

P